MODEL ANALYSIS OF STRUCTURES

MODEL ANALYSIS
OF
STRUCTURES

by

HEINZ HOSSDORF

Translated
by

C. VAN AMERONGEN

 VAN   NOSTRAND   REINHOLD   COMPANY

New York   Cincinnati   Toronto   London   Melbourne

© Bauverlag GmbH, Wiesbaden und Berlin, 1971
© English translation, Van Nostrand Reinhold Company Ltd., 1974

Published by Van Nostrand Reinhold Company Ltd.,
Molly Millar's Lane, Wokingham, Berkshire

INTERNATIONAL OFFICES
New York   Cincinnati   Toronto   Melbourne

ISBN 0 442 30018 2

Library of Congress Catalog Card No. 73-11581

PREFACE

The science of structural model analysis and testing is undergoing a decisive metamorphosis. Until fairly recently there was a tendency to look upon it as as a rather unacademic activity, more suited to the technician than to the theoretically-minded engineer. With its undeniable shortcomings it was thus regarded as something of a last resort to which one turned only if all other means of determining the behaviour of complex structures failed to provide the answer. But now all that is changing rapidly. In close association with the digital computer, the testing and analysis of structural engineering models is emerging as an indispensable tool for regular use by the practising designer.

The future of model analysis is closely bound up with the full utilisation of data-processing facilities. In addition, if advantage is taken of the other vast possibilities of modern electronics, more powerful and efficient methods of structural design than those attainable with the potentialities of the computer alone should be possible. The main object of this book is to outline this novel (and perhaps to some readers, surprising) process of development, together with some of the more traditional features of structural model analysis.

One of the objects of this book is to fill a gap in the existing literature. What is currently happening in the field of model analysis and testing techniques is based on the earlier work of many pioneers in this field. No single publication has so far attempted to give a representative selection from the many instructive examples of recent and early model experiments in testing laboratories around the world. The examples highlight the numerous and varied possibilities of application of structural model experiments and reveal something of the ingenuity and imagination brought to bear in solving the problems. Many of these examples, with their clarity and their direct appeal to one's visual powers, should be of interest to the architect and to the interested layman as well as to the structural engineer.

Basle, April 1971   H. Hossdorf

CONTENTS

3   EXPERIMENTAL TECHNIQUE

1.

GENERAL

1.1 WHY USE MODEL ANALYSIS?

The significance of structural models and the reasons for their use in stress analysis may be simply explained: their purpose is to provide the structural designer with a scientific technique to liberate him from the straitjacket of limited theoretical knowledge about the behaviour of structures, and to enable him to extend his design activities into the large unexplored field of physically practicable structures.

One would suppose, therefore, that this promising aid to the designer deserves all possible encouragement. Yet, from some engineers, the development of model analysis meets with scepticism. It will perhaps be useful to examine more closely the underlying cause for such unappreciative attitudes.

In the first place, there is undoubtedly what might be called a philosophical reason—a reason linked to the mental attitude towards the method of acquiring knowledge. Among philosophers there have always been rationalists, i.e. people who try to show that the truth about the fundamental laws of the physical world is already present *a priori* in the human mind. They feel that, in order to find out about the laws of Nature, it is merely necessary to explore the mind itself, without troubling to observe the world around us. The great German philosopher Immanuel Kant, perhaps the most famous representative of this school of thought, postulated that space comprises three dimensions, that two points can be connected by only one straight line and that the quantity of matter in the natural universe can neither increase nor decrease, whatever changes may occur in the phenomena. However, in terms of the modern scientific view of the universe, none of Kant's 'pure truths' which he supposed to exist *a priori* in the mind are acceptable today.

Modern physics has drawn its conclusions from the realisation of the impossibility of having *a priori* knowledge; it treats every model concept of natural processes, however obvious and convincing, as merely a working hypothesis. It is prepared to modify this hypothesis whenever fresh discoveries demand. However, this must not be taken as a plea for the empirical approach. It means that, in the light of compelling experience, Man must be ready to concede that he possesses no absolute cognitive capacity and to realise that his mind is competent to perform only one function: to order the available data, i.e. to marshal and co-ordinate them systematically.

We feel sad or humble at finding that an order based on theoretical concepts and analyses, when applied to a specific problem, often proves imperfect. It is

only in places where it is still possible to dodge the real problems—e.g. in the older and more conventional engineering schools, where 19th century mental attitudes linger—that these utopian ideas are still cultivated. Engineers graduating from such schools start their careers with the erroneous notion that all the problems they encounter must be capable of analytical solution. These graduates have been shamefully uninformed about the limitations of theories. Many of them cling to such ideas throughout their professional lives, more particularly those who spend their days producing uncreative designs which are mere reproductions of conventional structures.

This overvaluing of the analytical approach in engineering studies—now fortunately giving way to a different outlook—has another effect: the magical belief in numbers. To this sort of engineer the model test is unfamiliar, in that results are presented in analog form and not as definitive measurements. The numerical data obtained in this way suffer from a certain amount of scatter; they are affected by errors. A purely analytical calculation can be contrived to give results to any desired number of decimal places. Thus it possesses what appears to be a reassuringly high degree of accuracy. Familiarity with a theory which he continues to apply repeatedly is liable to make an engineer forget how imperfectly the assumptions on which he bases his calculations represent the true behaviour of the structure. The passion of such engineers for spurious accuracy is greatly indulged by the electronic computer!

Besides these quite unjustified 'reasons', there are sounder arguments, deserving serious attention, against the usefulness of model tests in the form in which they are generally employed at present; considerations which relate to the credibility of the experimental procedures themselves. Misgivings on this score are directed more particularly at the following points:

(a) The accuracy of measurements performed on models often leaves much to be desired.

(b) The construction of the model, the test set-up and the interpretation of the measurements are so time-consuming that in practice the execution of tests is often ruled out because of insufficient time.

(c) The cost of model tests for ascertaining precise structural behaviour is not economically justified, except for very large structures.

It is perhaps true that, so far, tests performed on models have often been inadequate rather than instructive. This is not due to the inherent nature of model testing, but solely to the fact that model-testing techniques are only now taking their first serious steps towards emancipation. Electrical methods of measurement have been perfected in recent years, and the use of computers for controlling the measuring sequences and for data processing has transformed the whole pattern of model testing, hitherto characterised by largely manual and very laborious methods. One of the principal objects of this book is to reveal the new possibilities that current computer technology and measuring techniques open up for the future development of model testing. It will be shown that, in principle, the engineering model of the future will be no more than an analog information carrier for the digital computer (Section 4.2).

Engineers of the future will have the choice of three different tools for solving structural problems:

(i) Conventional analysis with slide-rule and desk calculating machine.

(ii) The program-controlled computer.

(iii) Modern model analysis and testing.

What are the fields in which one or other of these three methods can be employed? The accompanying diagram (Fig. 1.1) attempts to give a qualitative demarcation of the respective fields of application.

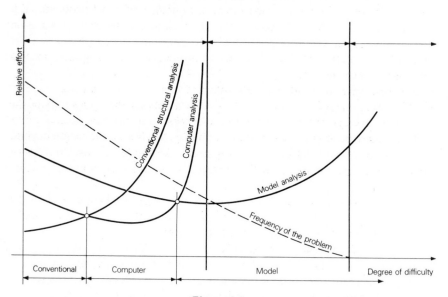

Figure 1.1

The relative amount of effort involved in solving a structural problem is adopted as the criterion and is plotted as a function of the degree of difficulty of the problem. The latter could, for example, be costed in terms of man-hours and hence of fees.

Two qualitative boundaries can be established on the horizontal axis. The first 'boundary of difficulty' is determined by the limitations of our *theoretical* knowledge of the load-carrying behaviour of structures and materials; the second is the absolute limit of the *physically* possible, beyond which the engineer can conceive but not actually construct.

Conventional methods of structural analysis, using slide-rule and desk calculating machine, can cope with simple problems. The effort increases steadily with the difficulty of the problems, and for the more complex ones encountered in structural theory will rapidly exceed the limit of what can reasonably be attempted.

The referral of isolated simple problems to a computing centre is pointless on account of the bother, time and cost it involves. The purchase of computers for private use by consulting firms and the establishment of long-distance

links with computer centres are factors which are bound to shift the situation in favour of the computer. For frequently recurring problems of moderate difficulty, the programmed electronic computer will often produce the desired results more speedily, conveniently and rationally than conventional methods. Big computers are also suitable, within limits, for solving relatively complex problems of elastic theory. It remains to be seen to what extent the computer programs that are being actively developed all over the world will be introduced into engineering design practice. With a powerful computer it is already possible to deal economically with a wide range of difficult analytical problems. But as soon as the problem approaches the first boundary, the effort of solution rises steeply along the asymptote that it has in common with the curve for conventional methods. Thus even the computer, and its programs, meet a boundary which they cannot cross (Section 4.1).

It is here that the fundamentally different character and unique value of model testing emerge. Its curve in the diagram cuts off the upper part of the computer curve, i.e. the part where the use of computers becomes uneconomic, and continues, beyond the asymptotic boundary, to open up the region beyond for quantitative exploration by the engineer. The model-testing curve stops only at the second boundary which is dictated by the absolute limit that Nature imposes.

In other words, model analysis and testing techniques liberate us from the fetters of our inevitably imperfect conception of Nature and provide a realistic picture of the behaviour of any type of structure.

We may ask whether this striving to achieve liberation reflects a real need—more particularly, if such liberation makes it possible to build complex structures. Is there any purpose in building structures which are characterised by fanciful shapes and which do not permit ordinary analytical treatment?

We live in a world in which most of the human race seeks its happiness in achieving ever greater utilisation of material resources with the least possible effort. Our social and economic life is largely derived from, and dictated by, this principle. In building construction, therefore, rationalisation and cost reduction represent the ultimate goal. The importance attached to prefabrication clearly reflects this state of affairs. In this area, there is as yet little scope or need for model analysis.

However, side by side with developments in constructional technology, there is another, less obvious and less spectacular, but no less important, trend. Man does not live by rationalisation alone. If that were so, we should never build theatres, museums or churches. Our need for art and the conditions of human life that go with it would have to be sacrificed to a system of social mechanics. But architecture still ranks as an art and, fortunately, we are still far from seriously desiring such a situation.

Together with rational attempts to produce low-cost utilitarian buildings, there is also the desire to make technology subject to laws other than its own. In good architecture, it is Man himself who determines the function that a building fulfils. The architect's creative will gives form and substance to the enveloping structure. Architecture also has to satisfy a number of other,

often complex, requirements such as acoustics, lighting, heating and ventilation. The creative and unprejudiced co-operation of the structural engineer is indispensable in producing a structure that will completely meet the functional demands made upon it. Even in the domain of pure construction technology, it is accepted that the statically simplest or cheapest structure will not necessarily best fulfil the functional requirements. The shape of a modern highway bridge is only partly decided by convenience of structural analysis; it is largely dependent on such considerations as efficient road alignment and aesthetic design, and proper fulfilment of these requirements often makes greater intellectual demands upon the designer. A real escape from stereotyped constructional forms is probably beyond our reach without the aid of model analysis.

In this context another, more general, aspect of model analysis emerges: the investigation of structural models as a catalyst between engineer and architect.

The cat and dog relationship between these two professions, which by the very nature of their relative functions should work together harmoniously, is hard to understand. Creative co-operation and mutual understanding and confidence between engineer and architect are, unfortunately, the exception rather than the rule.

Inspection of Figure 1.1 reveals the reason for this lack of harmony. Here again the origins can be traced back to the university. The engineering student is taught some of the things that can be done in the region situated to the left of the first boundary in that diagram. With analytical methods he makes his way slowly, to the best of his ability, towards this boundary (the limit of what is 'analytically definable'). He acquires the conviction that he can never reach it let alone cross it. He learns to work with his analytical techniques and the region beyond the boundary remains literally out of bounds to him while he is at his engineering school (where he is taught little about design). It is from this angle that he takes a disapproving view of architects who appear to be cheerfully, almost irresponsibly, playing about on the other side of his boundary without a thought for the technical difficulties that obstruct their ambitious ideas. It is a good and necessary thing that the architect revels in his freedom. He must not allow himself to be hemmed in by the limitations of structural analysis; he must not over-rate its importance. But he should be trained to sense and recognise the existence of the second boundary, which is indeed an absolute one. But how can we really expect the architect to judge this if the engineer's thinking already stops short at the first boundary? Engineering students should be taught more about structural design and given a greater insight into the scientific methods that help them test the feasibility of their own ideas. It is up to the engineer to make a creative contribution to architecture and to give guidance to the architect in seeking fulfilment of his constructional aims. Model analysis and testing could be the crucible where architect and engineer, working together, jointly evolve their ideas. During such co-operation the model can give the engineer results to evaluate scientifically and can provide the architect with a means of visualising the structure he wants to create.

1.2 MODEL TESTING AND ITS APPLICATION

Model analysis is envisaged as the whole set of procedures involved in deriving data design from structural models, comprising many different forms of experimental activity. The term 'model testing' reflects the practical aspects of 'model analysis'. No attempt will be made here to give a comprehensive general definition. Instead, we shall try to present a systematic account of the various types of model test. For this purpose it is helpful to consider tests from two different points of view; namely (i) the experimental technique employed and (ii) the purpose of the test.

1.2.1 CLASSIFICATION OF EXPERIMENTAL METHODS

A model is a kind of copy. To understand the true behaviour of a physical entity we have, besides direct observation, two means at our disposal: experiment and theory. A theory is merely a model constructed in the mind, an imaginary principle superimposed upon a given set of phenomena. It enables us to 'comprehend' a particular sequence of natural phenomena as a causal and logical process involving well-defined quantities. Any description of the elements of a mental model must therefore be incomplete and thus only be capable of giving an accurate picture of natural processes in a limited range of cases. Theories of structural analysis, a branch of applied physics, are also subject to this limitation.

The possible relationships between a real structure and the descriptive possibilities in the domain of stress analysis (strength of materials) are schematically represented in Figure 1.2 where the various conceivable types of model tests are indicated.

Reality (the prototype) can be simulated either by pure *theory* (the mental model) or by an *experiment* (the physical model) which seeks to achieve a copy of the prototype. Ideally, the model should be perfect in the sense of enabling general inferences to be drawn concerning the problem under investigation. Unfortunately, from experience, we know that accuracy and generality can never be fulfilled simultaneously. The closer and more realistic

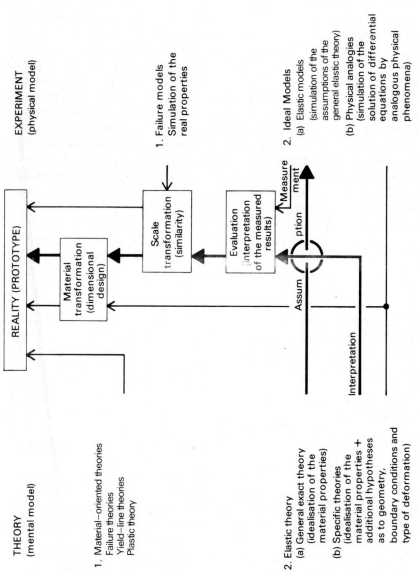

Figure 1.2 Schematic representation of the various theoretical and experimental possibilities for problems in stress analysis.

the simulation of physical behaviour, the less readily can we apply the knowledge thus gained to other cases. On the other hand, with ideal assumptions it is possible to find general (but imperfect) laws and relationships. This principle applies quite generally to theory as well as to the physical experiment. Thus we can control the degree of approximation to reality and so establish theories or perform experiments to provide either precise or more general information.

In theoretical stress analysis, the primary assumptions include material properties. The range of materials available such as metals, concrete or plastics provides a variety of physical properties and it seems logical to form separate laws for the behaviour of structures made of each material or combination of materials. This is reflected in the failure theories for reinforced concrete or pre-stressed concrete sections (e.g. yield-line theory) in the plastic theory of steel structures, or wherever an attempt is made to simulate the behaviour of structures through the whole range of loading up to failure. For individual cases, observation of actual failure can give a reliable picture of structural behaviour and provide a basis for the assumptions underlying an experimental study. However the conclusions which are of value for judging individual cases should not be generalised, because for one thing the complex failure mechanism is seldom fully understood and because the experimental result depends upon the arbitrarily chosen load distribution producing failure. Similarly, the general validity of yield-line theory suffers from the arbitrary nature of assumptions which cannot, in the last resort, stand up to objective assessment. Failure tests and failure theories are suited to the investigation of individual problems only if the load distribution is known.

Elastic theory adopts a fundamentally different approach to the behaviour of structures. The designation 'strength of materials' as applied to problems in elastic theory is misleading, since in actual fact elastic theory is not fundamentally concerned with the strength of the material in question. On the contrary, it endeavours to state the relationships between external loading and deformation, and between stress and strain, in a general analytical form. If the problem is to be solved mathematically, the material properties on which the theory is based have to be greatly simplified—so that failure (or ultimate strength) is not properly defined theoretically. However, under relatively small loads, most construction materials can be correctly described in terms of elastic theory. The great advantage of this theory lies in its general validity. It remains the way to arrive at a coherent understanding of the load-bearing behaviour of structures. The numerical data deduced from the theory must, however, be supplemented by the introduction of the actual material properties in order to determine the cross-sectional dimensions, etc.

Elastic theory is based on the familiar concepts of material properties (Section 2.2). Furthermore, to set limits on the mathematical complexities involved calls for additional simplifying hypotheses, e.g. kinematic ones, for describing special structural forms (Section 2.3), so that most solutions of elastic problems are at variance with the general assumptions. In practice, before undertaking the analysis of a special problem we must critically assess the assumptions to be made and decide on one or other theoretical approach.

In the experimental field, the procedure of the elastic-model test is the counterpart of elastic theory in its assumptions, in its general applicability and in its limitations when describing the phenomena of structural failure.

The technique and the philosophy of elastic-model testing are aimed at making and testing structural models whose properties, within the chosen range of observation, simulate as ideally as possible the assumptions and conditions of general elastic theory. If this succeeds, two decisive advantages over the theoretical approach are gained. The behaviour of the model simulates the solution of the theoretical problem without complicated differential equations. The model integrates the equations even under circumstances where a mathematical solution would not be practicable.

Approximate hypotheses that have to be formed to deal with special problems in elastic theory (e.g. Navier's hypothesis) cannot be simulated in the elastic model. The measured solution is *exact* by nature. This is in itself an excellent feature of the model test; it can, however, lead to difficulties of interpretation of measured data and the experimenter should always consider carefully the causes of such difficulties.

Individual measurements are seldom of much interest to the engineer. For the design of a structure he needs to know the stress resultants, and in order to calculate these from sets of measured data he must fall back upon theory. To interpret the measured data, the investigator will generally use the simplified relationships of elastic theory based on hypotheses to which the model does not strictly conform. In the schematic diagram (Fig. 1.2) this theoretical contradiction is shown by the intersection marked by a circle. The choice of the interpretative theory and the positioning of the measuring stations must be carefully considered and adjusted if there are not to be contradictions between measurement and interpretation. The processing and interpretation of results (Section 2.3) are therefore highly instructive in connection with understanding structural problems and their solution on the basis of elastic theory.

The results thus obtained must next, with the aid of dimensional analysis (Section 2.1), be transformed to the scale of the prototype and, like any result obtained from elastic theory, must be re-interpreted in terms of the material used in the actual structure to arrive at a design solution (Section 4.2). Elastic model testing necessitates better understanding in a number of areas:

(i) Simulation of the conditions of elastic theory by developing suitable model materials.

(ii) Further development and improvement of model testing and instrumentation to obtain exact information concerning the behaviour of the model.

(iii) Application of sections of general elastic theory to the interpretation of data; development of interpretation routines.

(iv) Knowledge of model mechanics.

(v) Further development of design methods based on the results of elastic-model tests; at present these methods leave much to be desired.

Finally, another technical factor in model testing may be pointed out. Although the model simulates idealised material behaviour and ignores the material of the prototype, even elastic models are structures with dimensions

which are similar or affine with respect to the prototype. They are supported and loaded in the same manner as the latter.

The abstraction of the model can be carried one major step further. In physics we often encounter analogously constructed differential equations which describe phenomena belonging to entirely different branches of science. Potential equations serve to describe hydrodynamic and aerodynamic flow conditions, and also electrical or gravitational fields. In all such cases it is possible, in theory, to investigate the behaviour of any particular phenomenon by means of an experiment on an analogous phenomenon in a different scientific discipline. The investigator can thus choose the method best suited to the needs of experimental technique. Electrical analogies, if they exist in any particular case, are technically the most convenient to manipulate.

Familiar analogies in stress analysis are Prandtl's membrane analogy for Saint Venant torsion and the representation of Bredt's formula by a hydrodynamic analogy.

The technique of solving differential equations by electrical analogies and of carrying out other mathematical operations with electrical circuits is consistently utilised in analog computers. These devices are able to perform linear operations such as addition, subtraction, integration and differentiation, and non-linear ones such as multiplication and division. Functions of any kind can be simulated. Analog computers are very suitable for the rapid solution of linear and non-linear differential equations, but less efficient in dealing with sets of algebraic equations. For this work, the digital computer is far superior. Analog and digital computers are now often combined into compatible systems, so-called hybrid computers, in which the digital computer performs the arithmetic operations and is programmed to monitor the whole computation process and to control the functions of the analog computer.

The analog computer has been used only to a limited extent for the solution of stress-analysis problems. Perhaps this is mainly because the structural engineer is still unfamiliar with its possibilities, and indeed its very existence, as a tool to aid him. Exceptions are the analog computer which has been used at the California Institute of Technology for the analysis of aerofoils (aircraft wings) and the analog computer designed to solve biharmonic problems at the Building Research Institute at Bratislava.

Analog computers and special hybrid computers are undoubtedly able to solve a number of other structural analysis problems. In this context the simulation of 'plastic design' methods and the treatment of higher-order problems come particularly to mind.

Although analog simulations in the sense referred to above cannot be regarded as true model tests, the mode of functioning of hybrid computers is important with regard to model-testing techniques. As in hybrid computer systems, the programmed digital computer will control the experimental sequence of future model tests and carry out the data processing to give final results (Section 4.2). The only fundamental difference will be that in automated model tests the measured data directly represent strains occurring in an actual model structure. The digital computers designed for controlling hybrid

computers contain all the logic elements and electronic components necessary for the elegant control of model tests. This much can be learnt from hybrid computers.

1.2.2 METHODS AND PURPOSES OF MODEL TESTING

As already stated, model tests can alternatively be considered from the viewpoint of the original purpose of the test. This approach to the question of classification provides an indication of the scope and range of application.

1.2.2.1 Qualitative tests

For a preliminary assessment of the properties of a new structural shape, it is common practice to make use of models made of simple materials such as paper, cardboard, wire, etc. They are loaded by simple manipulation in order to judge the structural behaviour. Experiments of this sort have the advantage of being highly adaptable as the model is modified quickly and easily in the light of any fresh information that emerges. In this way it is possible to make valid and reliable comparisons of the structural capacity of related structures. The fundamental behaviour of different structural systems can effectively be found by variation of the support conditions in the same model.

Critical assessment of such improvised experiments requires some knowledge of the principles of structural analysis, and more particularly of model mechanics, since small models can too easily give rise to misleading conclusions regarding the behaviour of the structure (Section 2.1). Qualitative tests can be a boon to the designer, but they can never provide conclusive proof of the suitability of a structure.

1.2.2.2 Semi-analytical model

For studying the behaviour of structures as a whole, it is often necessary to investigate the interaction of a variety of structural members. The structure may be composed of easily analysed sub-assemblies such as lattice girders, portal frames, etc., and also include complex plate or shell elements which are not properly amenable to analysis. In an elastic investigation of the overall problem, it is always possible, by separation of the 'difficult' components from the system as a whole, to determine their structural behaviour by model testing and thus find the deformations which occur at the joints with the rest of the structure when unit loads are applied. These experimentally determined values can then be introduced in the usual way into the equations for the structure as a whole. The final stresses in the experimentally tested sub-assembly are next

determined by evaluation of the measured influence coefficients for the joints—at which the cuts were applied to detach the sub-assembly from the rest of the structure—by introducing the calculated redundant quantities.

The principle of semi-analytical methods can be extended. The finite element method (Section 4.1) for the analysis of complex elastic structures has rapidly come into prominence in recent years. It is eminently suitable for use with computers and is based on the idea of considering the structure under investigation—which may be of any shape—as composed of a number of elements of finite dimensions whose elastic behaviour can be defined. The elastic properties of this assembly of elements (e.g. rectangular or triangular plates, cylindrical or spherical shell elements) are embodied in a stiffness matrix. The accuracy attained by this method will evidently approach the behaviour of the actual structure as a continuum as the chosen mesh of elements becomes finer, but this of course makes the analysis more laborious and may take up more computer time than is economically or indeed technically warranted.

On the other hand a coarser element mesh may be chosen, without affecting the final result when it is able to represent accurately the geometry of the actual structure. Hence, there is a need for the completest possible collection of elements with widely varying geometric features whose stiffness matrix is known. Here elastic experimentation can again intervene decisively by establishing sets of experimentally determined data.

1.2.2.3   The model as an adjunct to structural calculations

In the preceding section we considered the possibility of investigating entire structural assemblies elastically and of introducing the measured deformations per unit load into the analysis of the structure as a whole. Quite often, however, the latter can be analysed with sufficient accuracy in the usual way even if the state of stress at one or more 'sensitive' points is not precisely known. This is always so when obscure local states of stress are of minor importance to the general deformation and behaviour of the structure. In such circumstances the structure can be analysed and the forces determined can then be applied as test loads to the local element or part of the structure we wish to investigate. The following are a few examples: the stress distribution around a hole or cavity, the stress distribution in regions where concentrated loads are applied to a structure (e.g. bearings, end anchorages of pre-stressing cables, point loads) and the stress distribution in plates which function as rigid members in relation to the main structural system (e.g. diaphragms in bridge super-structures).

In many cases it is unnecessary to perform a full model test; it will suffice to measure a few data which will enable the investigator to pinpoint the essentials of a structural behaviour pattern that would otherwise be difficult to visualise. In such instances the simple measurement of bearing reactions proves most effective. If the reactions are known, it is often possible to achieve a complete understanding of highly redundant systems from simple considerations of

equilibrium. Other examples of this method of utilising model-testing techniques are to be found in illustrations (Section 3.2).

1.2.2.4 The model as an independent tool

The possibilities of the application of structural models are thus seen to be numerous and varied. The ultimate aim of model analysis will, however, always reside in its use as an independent research tool for analysing the structural behaviour of the system under investigation. The model should therefore provide a complete substitute for structural calculations, particularly in cases where such calculations are impracticable. With this in view, the great advantages of the elastic test become manifest. It retains the versatility and adaptability that characterise elastic theory, in that the principle of superposition is valid also for this type of test. Structures of any shape, with any boundary conditions and any type of loading (including pre-stress, settlement of supports, temperature effects, etc.) can be accurately determined within the range of elastic behaviour. The fact that model testing is still only used with hesitation is due *solely* to the lack of development of the testing technique. As explained in Section 4.2, these difficulties are now certain to be speedily overcome. Model analysis will become an indispensable tool for the engineering designer to which he will have recourse as readily and conveniently as to the computer.

1.3 HISTORICAL DIGRESSION

The idea of model testing is as old as Man's desire to comprehend natural phenomena. We give our children dolls or boxes of bricks to play with because of our instinctive conviction that while playing with reduced-scale models they will prepare themselves for their subsequent encounter with the real world. All laboratory experiments, in whatever scientific field, are in a sense model tests. Indeed, in all fields of technology, investigators adopt analog experiments as an indispensable means of acquiring knowledge. In view of this, the scepticism of some structural engineers towards model testing—as referred to in Section 1.1—hardly deserves to be taken seriously. Only a lack of realism permits withdrawal into such an ivory tower.

It is not very surprising that Leonardo da Vinci was already aware of the existence of model laws and mathematical relationships for the modes of behaviour of geometrically similar structures. On the basis of similarity, he established a 'comparative' theory for a few simple constructional shapes.

The testing of models in structural engineering and building construction was not much practised until well into the first quarter of the present century. The reason must be sought in the fact that the evolution of architectural forms has thousands of years of tradition and therefore has proceeded at a much slower rate than, for example, comparable developments in mechanical engineering. Yet the builders of the Gothic cathedrals of the Middle Ages are known to have used models for investigating the stability of arch structures. Until the latter part of the 19th century, almost the only basic structural

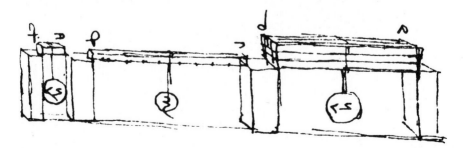

Figure 1.3 Leonardo da Vinci: sketch intended to illustrate the principle of elasticity (Cod. Atl., folio 152 recto-b).

Figure 1.4 Leonardo da Vinci: study of similarity relationships (Cod. Atl., folio recto-b).

members at the engineer's disposal, besides the vault, were the beam, the arch and the lattice truss. Accordingly, engineering was very largely concerned with this very limited range of structural shapes which could be realised in timber or iron (later steel). An exception to this rule was formed by some iron bridges built in Britain—in those days, the most technologically advanced country—such as the Britannia bridge, a tubular railway bridge over the Menai Straits, whose design embodied features which even now have a 'modern flavour' (box girders constructed from riveted wrought-iron plates and subject to a combination of flexural, torsional and buckling load conditions). In those days (before 1850), nothing was known about the buckling stability of such girders. Robert Stephenson and William Fairbairn, the designers of the Britannia bridge, carried out tests on a 1:6 scale model of this structure. That was in 1846. Model tests were also undertaken for the design of the Conway bridge, another tubular structure for railway traffic. Again in Britain, Telford is known to have used models for testing the behaviour of suspension bridges in the early part of the last century.

Up to the end of the 18th century, timber was the material which, because of its elasticity, workability and load capacity, afforded most scope for the engineering designer's creative imagination. So it is hardly surprising that some of the greatest structural engineers of those days had been trained as carpenters by trade. The well-known story of the construction of the bridge over the Rhine at Schaffhausen in Germany strikingly illustrates the approach to engineering at that time. It was in 1755 that Grubenmann, a carpenter, proposed building a timber bridge over the river and in due course was commissioned to carry out the work after having demonstrated the soundness of his design with the aid of a model.

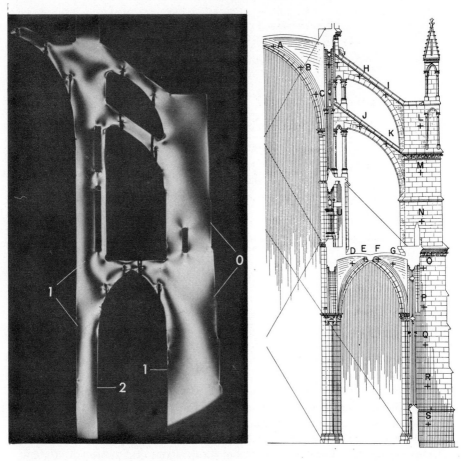

Figures 1.5 and 1.6   Section through part of Amiens cathedral, a mediaeval Gothic construction, and a photoelastic model made of CR-39 under loading due to dead weight. This was one of the tests carried out by Robert Mark at Princeton University, U.S.A. for verifying the structural behaviour of some Gothic buildings in France. Weak points in the structure were shown up by the model tests. Inspection of the actual buildings revealed the presence of cracks in the masonry at the predicted points.

Perhaps not so well known, but nevertheless significant in connection with assessing the merit of his work, is the fact that the celebrated and controversial Spanish architect Antonio Gaudi also made use of model tests in designing his romantically conceived 'spatial' constructional shapes (more particularly in the period 1900–1910). With the aid of 'inverted vaults'—cable structures from which appropriate weights were suspended—he sought the correct shape which, once established, he consistently applied to actual structures. His buildings display a freedom in the treatment of shapes unusual in masonry construction and is often, in a superficial judgment, erroneously rated as merely fanciful. In actual fact, Gaudi's structures embody a high degree of technical functionalism.

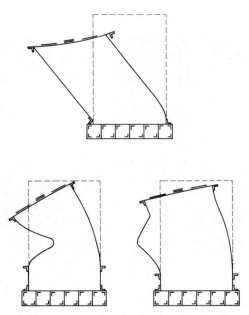

Figure 1.7   Results of experiments performed on 1:6 scale models by Fairbairn in 1864.

The origins of scientific model testing are to be found chiefly in the inter-war years. It is perhaps significant that this was also the period when reinforced concrete came into its own as a building material that revolutionised architecture. Having begun as a substitute for timber, stone or steel, it evolved into a material with its own laws which are, above all, strikingly manifested in the varied structural shapes we see today. As a result of this new freedom in terms of design possibilities, structural engineers soon broke through the existing barrier, formed by the limit of 'analysable' engineering structures (relating essentially to assemblies of beam- and strut-type members conforming to conventional flexural theory). Particularly in Germany, structural analysts set to work to make plate and shell structures*, now gaining rapidly in importance, amenable to calculation. To build anything without a basis of structural analysis—i.e., design calculations in the traditionally accepted sense—was unthinkable in Germany in those days. Names like Dischinger, Flügge and others are closely associated with these developments. They established a mathematical basis for calculating some of the simpler shell roof shapes and thereby helped the new constructional medium—reinforced concrete—to achieve a major breakthrough.

Increasing freedom of structural forms was accompanied by a vast increase in the numerical computation required. This quickly outgrew its original purpose, namely to furnish analytical proof of the strength, i.e. the load-carrying capacity, of a structure. It was therefore imperative to seek some other

* To some extent identifiable with 'stressed-skin structures', a term borrowed from aeronautical engineering and implying that the surface material performs a structural function in its own right and is not just a weather-excluding cladding affixed to load-bearing members.

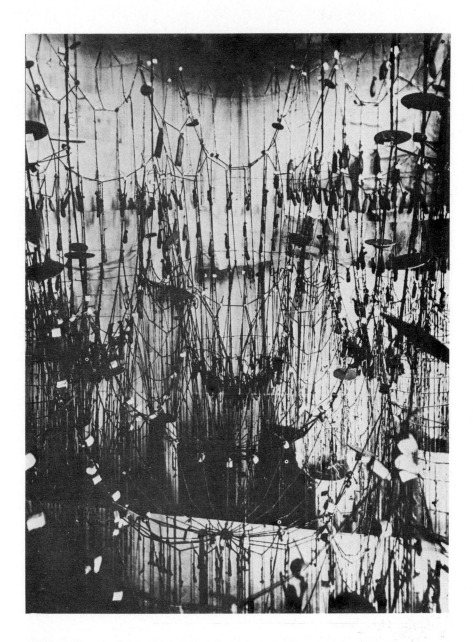

Figure 1.8   In 1905 Antonio Gaudi determined the correct structural shape for a church (Iglesia de la Colonia Güell) at Barcelona by testing a model (shown here loaded with weights attached to it by cords and representing the dead weight of the masonry). The supporting system is designed to function as an inverted arch.

Figures 1.9 and 1.10 Views of the structural masonry of the Colonia Güell church.

method of understanding and estimating quantitatively the behaviour of these new structural shapes.

Besides promoting shell roof construction, the development of these advanced analytical methods of design had certain restrictive effects on the shaping of such structures. Obviously, present-day computer methods are potentially prone to the same hazard. The 'dismemberment' of the structure that the

analytical approach requires often compels the designer to adopt an artificially contrived geometric configuration to adapt it to his particular method of analysis. This need is in evident contradiction to the designer's ultimate aim: freedom in determining the shape of structures. A typical example of this is afforded by barrel-vault shell roofs with 'edge members', of which thousands have been built all over the world. The edge member in this form of construction was introduced purely to fulfil the requirements of conventional barrel-vault shell theory. Physically this edge member, in the obtrusive and unsightly form often adopted for it, can be dispensed with. Something similar occurred in the construction of ribbed domes. A comparison of the analytically calculated market hall roof at Basle with that at Algeciras, whose design was based on model tests, strikingly reveals the restricting effect of mathematical thinking upon the design of structural shapes.

The freedom in choosing whichever shape appears most appropriate to fulfil the needs of the structure is typically associated with reinforced concrete construction, but it has not been without its effect on steel construction and latterly also on construction with plastics. For example, the knowledge acquired of the structural properties and behaviour of plate and shell structures has also led to the development of modern welded plate structures in steel. It must not be overlooked, however, that the more recent progress in the theoretical and experimental design of stressed-skin structures has come largely from aeronautical engineering.

In this brief historical survey, the place of special importance accorded to Eduardo Torroja, the famous Spaniard, is not only because he can justly be regarded as the father of modern scientifically-conducted structural model

Figure 1.11   Eduardo Torroja.

testing, but also because of his whole mental attitude and the nature of his work as an engineer are in many respects quite exemplary. He was a creative designer who weighed the rational and artistic viewpoints against each other in a masterly manner. He was both theoretician and experimenter, using whichever means appeared best suited to the realisation of his constructional ideas. Torroja certainly did make use of structural analysis, but in his awareness of the shortcomings of the available mathematical resources he had to employ model testing as his tool. First and foremost comes the need to design an appropriate sound load-bearing structure which takes on form and substance as a result of rational considerations coupled with intuition tempered by experience. The question as to the checks for structural safety which must be carried out in the interests of public safety will have to remain a matter of secondary importance if unnecessary obstacles are not to be placed in the path of further development of building construction by excessive rules and regulations.

Besides Torroja, a number of other well-known engineers who were his contemporaries also made use of testing as a design tool. In assessing the merits of model testing, it is significant to note that the structures of those engineers who availed themselves of this technique at the time (up to the Second World War) are acknowledged as major innovators in the design of structural shapes. A feature that they all have in common is a highly developed sense of shape

Figure 1.12 Freely-shaped barrel vaults pierced by openings designed by Torroja. This roof structure for the Fronton Recoletos at Madrid, which in terms of daring and elegance of design has remained unrivalled in its class, was destroyed in the Spanish Civil War.

Figure 1.13 A model made of micro-concrete for verifying the validity of the elaborate design calculations for this roof. Simple but highly ingenious techniques were employed for the model test.

Figure 1.14 These grandstand roofs at Zarzuela racecourse are composed of parabolic hyperboloid shells designed by Torroja, one of the earliest examples of their kind (1936) and still modern-looking and aesthetically pleasing.

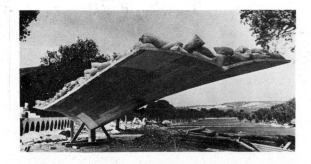

Figure 1.15   As with nearly all his advanced structural designs, Torroja performed model tests for determining the actual behaviour of these grandstand roofs; in this case a 1:1 scale model was tested.

Figure 1.16   Model for a yacht club house in Venezuela, designed by Torroja. Unfortunately, this graceful freely-shaped shell roof was never actually constructed.

and a scientific comprehension and deeply conceived intuitive grasp of the material properties, rather than any primary aesthetic consideration.

Robert Maillart, the renowned Swiss bridge builder, should also be mentioned here. He shocked petty officialdom by his habit of scribbling the design calculations for his world-famous reinforced concrete bridges on the backs of menus in restaurants. He also first introduced the mushroom floor or flat-slab floor and developed his designs for such structures from tests on reduced-scale concrete models.

Pier Luigi Nervi, the Italian architect and engineer, investigated the three-dimensional structural behaviour of his well-known aircraft hangars on models made of celluloid. At his initiative, model tests were also performed in connection with the design of skyscrapers, notably the Pirelli building at Milan.

Figure 1.17   Aircraft hangar at Orvieto, designed by Nervi (1936).

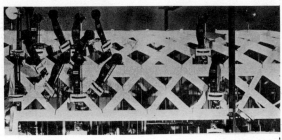

Figure 1.18   Close-up view of part of the model, showing the mechanical strain gauges attached. [Another structural model tested by Nervi is illustrated in Chapter 5 at the end of the book (Section 5.4).]

That Le Corbusier, the famous architect, also tested models for checking the feasibility of an idea for a factory roof structure is perhaps not so well known.

After the Second World War, model analysis and testing spread rapidly through every continent. Particularly in the East European countries, much effort was devoted to the further development of the procedures involved. The examples presented in this book are a very limited selection from the wide and varied range of activities conducted in laboratories for structural model research in many parts of the world.

It must be admitted that model-testing technique in its established form still leaves quite a lot to be desired as regards flexibility and adaptability. Nevertheless, we are entering a phase of new and revolutionary developments in experimental methods.

2.

THEORY

2.1 SIMILARITY MECHANICS

2.1.1 INTRODUCTION

Model mechanics is concerned with the extent to which inferences drawn from observations of physical phenomena on a mechanical system S are quantitatively applicable to an analytically similar system S' of different scale. The concept of similarity (or similitude) here relates not merely to the ratio of the geometrical dimensions of the two systems but also, in a general way, to comparable physical quantities such as time, force, acceleration, frequency, temperature, etc.

In all cases where the actual mechanical laws (i.e. the governing differential equations) for the phenomenon under consideration are known, these may be applied and transferred to model and prototype provided that the assumptions on which the derivation of the analytical relationship is based are equally valid for both. For a proper understanding of model mechanics and its meaningful application, it is important to bear in mind that mathematically formulated natural laws are valid only *within a limited range of scale* and give no reliable information on phenomena outside that range. When applying the laws of transfer associated with model mechanics, it must always be verified that the conditions and assumptions on which the mechanical laws applied are based are adequately fulfilled both for the prototype and for the model. If so, the similarity conditions must already be embodied in the differential equation.

As an example, the similarity conditions for a model test on a flexurally loaded elastic plate will be deduced from the differential equation for such a plate.

For the prototype the differential equation is of the following well-known form:

$$\Delta\Delta w = \frac{\partial^4 w}{\partial x^4} + \frac{\partial^4 w}{\partial x^2 \partial y^2} + \frac{\partial^4 w}{\partial y^4} = \frac{p}{K} = 12(1 - \mu^2)\frac{p}{Eh^3}.$$

We shall assume that an analogous differential equation holds true for the

model:

$$\Delta\Delta w' = \frac{\partial^4 w'}{\partial x'^4} + \frac{\partial^4 w'}{\partial x'^2\, \partial y'^2} + \frac{\partial^4 w'}{\partial y'^4} = \frac{p'}{K'} = 12(1 - \mu'^2)\frac{p'}{E'h'^3},$$

where the 'prime' denotes the quantities relating to the model.

We shall now investigate the necessary conditions for the existence of scale factors whereby the differential equations of the prototype and of the model can be transformed into each other.

$$w' = \lambda_w w \qquad x' = \lambda_l x \qquad y' = \lambda_l y \qquad p' = \lambda_p p \qquad E' = \lambda_E E$$

$$h' = \lambda_h h \qquad \mu' = \lambda_\mu \mu.$$

If these constants exist, the second equation can be written in the form:

$$\Delta\Delta w' = \frac{\lambda_w}{\lambda_l^4}\left(\frac{\partial^4 w}{\partial x^4} + \frac{\partial^4 w}{\partial x^2\, \partial y^2} + \frac{\partial^4 w}{\partial y^4}\right) = 12(1 - \lambda_\mu^2 \cdot \mu^2)\frac{\lambda_p \cdot p}{\lambda_E \lambda_h^3 E h^3}$$

To transform the model equation into the equation for the prototype, it is obviously necessary to satisfy the following conditions:

$$\frac{\lambda_w \lambda_h^3}{\lambda_l^4}\frac{\lambda_E}{\lambda_p} = 1 \tag{2.1}$$

and

$$\lambda_\mu = 1 \tag{2.2}$$

The second condition expresses the well-known fact that exact similarity of plates is attainable only if Poisson's ratio is the same for the model and the prototype.

The first equation gives the most general relation to be satisfied by the scales for deflection, plan dimensions, plate thickness, modulus of elasticity and loading. Geometrical similarity between model and prototype insofar as plate thickness is concerned is not essential to satisfy the conditions arising from the plate equation.

However, if complete geometrical similarity is provided and if it is desired that the relative deflections of model and prototype are equal, i.e. $\lambda_w = \lambda_h = \lambda_l$, then the condition is reduced to:

$$\lambda_E = \lambda_p \qquad \text{or} \qquad \frac{p'}{p} = \frac{E'}{E},$$

which means that, presupposing equal values of μ, the deformations of geometrically similar plates are dependent upon the geometrical similarity factor if the loading per unit area is proportional to the respective moduli of elasticity.

Transformation of the stress resultants measured on the model into those for the prototype can be deduced in an analogous manner from the corresponding differential equation with the aid of the general conditions expressed in equation (2.1)*.

---

*The term 'stress resultants' comprises the internal forces (shear force, normal force) and internal moment at a section of a structural member; each of these resultants is conceived as the integral of the respective set of stresses acting at that section, e.g. the normal force is the resultant integral of the normal (or direct) stresses.

For example, let the moment scale be $m' = \lambda_m \cdot m$. The internal moments as obtained from the curvature of the prototype are:

$$m_x = -\frac{Eh^3}{12(1 - \mu^2)}\left(\frac{\partial^2 w}{\partial x^2} + \mu\frac{\partial^2 w}{\partial y^2}\right); \qquad m_y = \cdots \qquad m_{xy} = \cdots$$

and for the model they are:

$$\lambda_m m_x = -\frac{\lambda_E \lambda_h^3 \lambda_w}{\lambda_l^2}\frac{Eh^3}{12(1 - \mu^2)}\left(\frac{\partial^2 w}{\partial x^2} + \frac{\partial^2 w}{\partial y^2}\right)$$

whence the model conditions are found:

$$\frac{\lambda_m \lambda_l^2}{\lambda_E \lambda_h^3 \lambda_w} = 1.$$

It can easily be shown that this transformation relationship has the same form for m_x and m_{xy}. The unknown scale λ_w can now be eliminated with the aid of the general condition given in equation (2.1). We thus obtain the model law for the internal moment, which is independent of the plate thickness and of the modulus of elasticity:

$$\frac{\lambda_m}{\lambda_l^2 \lambda_p} = 1 \qquad \text{or} \qquad m = m'\frac{p}{p'}\lambda_l^2.$$

As already stated, the range of validity of a model law (law of model similitude) derived from differential equations is confined to those physical phenomena for which the law was established. When applying model laws it is therefore necessary to make quite sure that the assumptions on which the derivation of the differential equation is based are indeed fulfilled both for the prototype and for the model. For example, the model law for tests on plates allows the plate thickness to be reduced and the loading to be increased to any desired extent. But it is obvious that a very thin plate under very heavy loading will deflect so much that membrane effects will considerably modify the load-bearing action of the plate in relation to the assumptions adopted in the derivation of the plate equation. The model law for membrane-type plates must therefore have a different form from that of equation (2.1) (Section 2.3.3).

Apart from the limited range of validity of model laws derived from differential equations, there exist in the domain of structural analysis, as in every other field of technology, many phenomena which are of practical importance and which cannot, or can only with great difficulty, be represented in the form of a differential equation. It is particularly in connection with tackling these phenomena that the model test is sometimes the only available means of obtaining precise information. Examples of such mathematically intractable problems are the aerodynamically generated oscillations of structures, the effects of variable temperature patterns, seismic effects, etc. In such cases it is virtually impossible to deduce model laws from analytical relationships.

It is possible to determine the exact transformation laws, i.e. the laws governing the extrapolation from model test to prototype, without knowing the precise construction of the differential equations involved. It is sufficient to identify all the physical quantities that influence a complex phenomenon. *Dimensional analysis* is concerned only with seeking to establish similarity laws through a knowledge of the physical quantities involved. In general it is a very useful instrument of research, whose importance extends far beyond model testing and analysis. It enables far-reaching comparisons to be made and thus establishment of working hypotheses for experimental research.

The hazards attending the uncritical application of dimensional analysis have already been noted. As we shall see, the application of dimensional analysis is based directly on the construction of our physical laws. It is implicitly assumed that for the physical phenomenon under observation there exists a law which, though not known to us in detail, can be described in terms of the physical quantities with which we are familiar. Since all 'natural laws' are in fact simplified model concepts constructed by the human mind on the basis of limited empirical data, they are valid only for a limited range of scale. So if transformation laws are derived purely through the medium of dimensional analysis, it is above all essential to satisfy ourselves that the physical quantities which we have introduced really are the decisive ones, both of the scale of the model and of that of the prototype. It may be necessary to introduce parameters which have no perceptible effect on the behaviour of the prototype and yet are of major importance to the behaviour of the model. Consider, for example, the predominant influence of surface tension on the shape of a drop of water and its virtual lack of influence on the shape of a wave at sea. In both cases we are dealing with water.

Dimensional analysis, therefore, does not determine the type or the number of necessary conditions—and still less the adequacy of such conditions—for establishing similarity laws. It does, however, supply the rules for obtaining a large selection of possible similarity conditions which would otherwise have to be found by mere chance. The choice of the physical parameters which are considered to be of decisive significance is left to the experimenter, and only his personal insight into their effects upon the phenomenon under investigation will help him to perform the test completely and conclusively. The possibility of carrying out comparison tests on models constructed to different scales can provide a means (though admittedly a laborious one) of verifying the correctness of the choice of the transformation parameters.

The uncertainties associated with the use of dimensional analysis, and the pitfalls that may be encountered, could at first sight raise doubts about the scientific reliability of the method. For this reason it must again be stated emphatically that as regards the accuracy and completeness of the model laws thus determined it makes no difference whether they have been derived from known physical laws (differential equations or special solutions thereof) or have been obtained with the aid of dimensional analysis. Although a differential equation in itself tells us nothing about its range of validity (and therefore about its applicability to different scales), we are, when using known differential

equations, better enabled by experience to assess the validity limits. In establishing differential equations for physical phenomena, it is, just as in applying dimensional analysis, always necessary to consider the intended range of application and the permissibility of simplifying assumptions. If 'ready-made' physical laws are used for deducing similarity conditions, the considerations applied in establishing these laws will have to be re-assessed with reference to the desired range of validity. Such re-assessment is entirely identical with the considerations that must precede any dimensional analysis. The latter, however, is much more generally applicable since it does not necessitate any knowledge of the physical laws themselves. For its application, it requires less analytical technique but, on the other hand, because of its much wider and freer scope, it calls more especially for sound judgment.

It is not always possible to infer the behaviour of the prototype from the results of measurements performed on models. Inadequacies and imperfections inherent in the experimental apparatus (e.g., wind tunnel tests in the sonic range) will not be considered in this context. But cases occur repeatedly (Section 2.2) where experimental results turn out to be different from those expected on the basis of the laws of similitude. Such phenomena are called *scale effects*. In such circumstances the investigator is (especially if the discrepancies are small ones) liable to dismiss the problem by assuming that the transformation laws were not quite accurate. He hopes that future results will be duly corrected with the aid of empirical correction factors deduced from repeated experiments. This approach may sometimes be justified for practical reasons but can hardly be regarded as satisfactory. The concept of 'scale effects' is vague and may be misleading, as it is liable to create the false impression that the model laws are inexact. Yet the scale in itself has no effect!

Instead of thus introducing a new concept to characterise the shortcomings encountered in effecting the transformation, i.e. the extrapolation from model to prototype, it would appear more suitable to distinguish clearly the possible reasons for the observed discrepancies. There are in fact only two such reasons: either the law of similitude used for making the comparison between model and prototype is *incomplete*, i.e. one or more physical parameters which exercise an influence on the behaviour of the model and/or the prototype have been overlooked to some measurable degree, or the physical conditions under which, according to the law of similitude, the model experiment ought to take place cannot be achieved for technical reasons.

2.1.2 HOMOGENEITY OF DIMENSIONS AND DIMENSIONAL SYSTEMS

Dimensional analysis starts from the familiar assumption that a generally valid physical law must be dimensionally homogeneous, i.e. in mathematical equations which describe a physical condition or process, the dimensions (products of powers of the basic units) must be equal for all the terms of a sum.

First, the following three concepts must be distinguished:

  (i) The physical quantity as a concept.

 (ii) The dimension applied to that quantity.

(iii) The dimensional unit, i.e. the unit of measure in which the dimension is expressed.

Physical quantities are more particularly those measurable operative quantities which are defined in physics for the purpose of the causal description of natural phenomena and which, because of their frequent occurrence, are regarded as sufficiently significant to be given a distinctive name. These include such quantities as velocity, moment, energy, electric charge, etc. The choice of physical quantities identified by their own names is fundamentally arbitrary. For example, the change of velocity with time is, for merely practical reasons, designated as acceleration. Acceleration itself may be subject to a time-dependent change but there is no special name for this.

A second group of quantities with which we are concerned comprises the *physical constants* which embody certain characteristic properties of the physical world in a quantitative manner. They are analytically linked to the operative quantities by equations of definition which characterise the experiment needed for determining them. The choice of recognised physical constants is also arbitrary and is based largely on practical experience.

Investigation reveals that it is possible to choose a limited selection of physical quantities as 'building blocks'. As dimensional units, they can be utilised for building up the dimension of any other physical quantity. This possibility follows automatically from the nature of the analytical constitution of our physical laws whose equations of definition, like those for the physical constants, always have the form of products of powers of the operative quantities.

If we take the structure of the analytical system of physics as our pattern and choose a sufficient number of basic quantities as dimensional units, all newly-defined operative quantities and constants can have dimensions which are products of the basic units in accordance with the prescribed laws defining the new quantities. Here again the choice of basic quantities is largely an arbitrary one and is governed by practical considerations. Thus we always choose basic units which will give the simplest possible expressions for the dimensions.

Dimensions

(a) *of operative quantities*	Mass system	Force system
Length	$[L]$	$[L]$
Time	$[T]$	$[T]$
Mass	$[M]$	$[FL^{-1}T^{2}]$
Force	$[MLT^{-2}]$	$[F]$
Temperature	$[\theta]$	$[\theta]$
Velocity	$[LT^{-1}]$	$[LT^{-1}]$
Acceleration	$[LT^{-2}]$	$[LT^{-2}]$
Angle	$[1]$	$[1]$

Dimensions

(a) *of operative quantities*	Mass system	Force system
Angular velocity	$[T^{-1}]$	$[T^{-1}]$
Angular acceleration	$[T^{-2}]$	$[T^{-2}]$
Pressure and stress	$[ML^{-1}T^{-2}]$	$[FL^{-2}]$
Moment	$[ML^2T^{-2}]$	$[FL]$
Energy, heat	$[ML^2T^{-2}]$	$[FL]$
Impulse	$[MLT^{-1}]$	$[FT]$
Power	$[ML^2T^{-3}]$	$[FLT^{-1}]$
Second moment of area	$[L^4]$	$[L^4]$
Mass moment of inertia	$[ML^2]$	$[FLT^2]$
Surface tension	$[MT^{-2}]$	$[FL^{-1}]$
Strain	$[1]$	$[1]$

(b) *of physical constants*		
Specific gravity	$[ML^{-2}T^{-2}]$	$[FL^{-3}]$
Density	$[ML^{-3}]$	$[FL^{-4}T^2]$
Modulus of elasticity	$[ML^{-1}T^{-2}]$	$[FL^{-2}]$
Poisson's ratio	$[1]$	$[1]$
Dynamic viscosity	$[ML^{-1}T^{-1}]$	$[FL^{-2}T]$
Kinematic viscosity	$[L^2T^{-1}]$	$[L^2T^{-1}]$
Linear coefficient of thermal expansion	$[\theta^{-1}]$	$[\theta^{-1}]$
Thermal conductivity	$[MLT^{-3}\theta^{-1}]$	$[FT^{-1}\theta^{-1}]$
Specific heat	$[L^2T^{-2}\theta^{-1}]$	$[L^2T^{-2}\theta^{-1}]$
Thermal capacity	$[ML^{-1}T^{-2}\theta^{-1}]$	$[FL^{-2}\theta^{-1}]$
Heat transfer coefficient	$[MT^{-3}\theta^{-1}]$	$[FL^{-1}T^{-1}\theta^{-1}]$

Depending on the choice of basic units adopted, different *systems of units* can be constructed and any number of such systems is conceivable. In practice a few have gained general acceptance, differing from one another in the degree of convenience they offer in their application in particular branches of science. Thus, in the so-called 'force system' the dimensions of the quantities occurring in statics can be constructed with two basic units, whereas the 'mass system' requires three of these for the same purpose. On the other hand, the latter system usually results in simpler dimensions when applied in general mechanics. In determining dimensional units, both take Newton's law as their starting point and differ only in that the first system takes force and the second takes mass as the fundamental 'building block'.

Other systems of units result in radically different dimensions. An example is afforded by the astronomical system in which the dimension of mass is deduced as follows from Newton's law of gravitation, adopting a gravitational constant $k = 1$:

$$[P] = [m \cdot b] = \left[\frac{mm'}{r^2}\right] \text{ or the dimensions } [MLT^{-2}] = [M^2L^{-2}]$$

Hence the dimension of mass is: $[M] = [L^3T^{-2}]$, and of force: $[F] = [L^4T^{-4}]$.

Hence all sorts of systems of units are conceivable without coming into conflict with the laws of physics.

In the 19th century there was considerable discussion among scientists as to the 'true' dimension of a physical quantity. Max Planck's observation should suffice to dispose of this controversy: 'To ask about the true dimension of a physical quantity is as pointless as to ask about the true name of an object.'

So, just like the names we give to things, the dimensions that we introduce are largely a matter of convention, though admittedly one which must be observed in considering a set of problems.

2.1.3 DIMENSIONAL ANALYSIS

We shall seek to combine the physical quantities considered to be significant for describing a phenomenon into dimensionless parameters. A useful aid in facilitating the search for such possible dimensionless parameters is the *dimensional matrix*. This matrix comprises a number of columns each representing a physical quantity, and as many rows as there are basic units necessary for establishing the dimension of the physical quantities concerned. Thus the elements of the matrix are the dimensional exponents of each particular physical quantity.

This procedure will be illustrated with reference to two examples. For a problem in elastic theory, the physical quantities E, l and μ for the structural member, and the quantities l, P, R, σ, δ and ε for the loads, reactions and deformations, are adopted as the significant ones. If the force system is chosen for determining the dimensions, the basic units L and F are sufficient, and the matrix of exponents will then be as follows:

	l	E	μ	P	R	σ	δ	ε
L	1	-2	0	0	0	-2	1	0
F	0	1	0	1	1	1	0	0

By comparison of the exponents in the columns we can, for example, readily write the following dimensionless parameters:

$$\pi_1 = \frac{\sigma}{E} \qquad \pi_2 = \frac{l}{\delta} \qquad \pi_3 = \frac{P}{R} \qquad \pi_4 = \mu \qquad \pi_5 = \varepsilon \qquad \pi_6 = \frac{P}{El^2}.$$

A number of other dimensionless parameters can also be found, e.g.:

$$\frac{\sigma l^2}{P}; \qquad \frac{P\sigma}{ER}; \qquad \frac{\mu}{\varepsilon} \text{ etc.}$$

However, all these additional parameters can be found by multiplication from the above dimensionless parameters. For instance, in the present example we may write:

$$\frac{\sigma l^2}{P} = \frac{\sigma}{E}\frac{El^2}{P} = \pi_1 \pi_6^{-1}.$$

In general, all the possible dimensionless parameters can be expressed in the following form:

$$\pi_n = \pi_1^{k_1} \cdot \pi_2^{k_2} \cdot \pi_3^{k_3} \cdot \pi_4^{k_4} \cdot \pi_5^{k_5} \cdot \pi_6^{k_6}$$

where $k_1 \cdots k_6$ are arbitrary exponents. The quantities $\pi_1 \cdots \pi_6$ are referred to as a *complete set of dimensionless parameters*, as each parameter in the set is independent of the others and each further possible dimensionless parameter can be derived as a product of powers of the quantities contained in the set.

Figure 2.1 Hovercraft model undergoing tests in a wave tank (Hovercraft Development Ltd.).

It can be shown that for eight physical quantities whose dimensions can be constructed from two basic units, a complete set of $8 - 2 = 6$ dimensionless parameters is determinable.

A second example of the procedure for establishing the dimensional matrix is taken from aero- and hydro-dynamics. The phenomena of fluid flow can be described with the aid of the physical quantities length l, pressure p, velocity v, density ρ, viscosity μ, gravitational acceleration g, velocity of sound c and surface tension σ.

Adopting the mass system, which is preferable in cases where dynamic phenomena have to be described, we obtain the following dimensional matrix:

	l	p	v	ρ	μ	g	c	σ
L	1	-1	1	-3	-1	1	1	0
T	0	-2	-1	0	-1	-2	-1	-2
M	0	1	0	1	1	0	0	1

A complete set of dimensionless parameters is given by the following combination of physical quantities

$$R = \frac{vlp}{\mu} \quad \text{Reynolds number}$$

$$P = \frac{p}{\rho v^2} \quad \text{pressure coefficient}$$

$$F = \frac{v^2}{lg} \quad \text{Froude number}$$

$$M = \frac{v}{c} \quad \text{Mach number}$$

$$W = \frac{\rho v^2 l}{\rho} \quad \text{Weber number}$$

Figure 2.2   Prototype of the Hovercraft (Hovercraft Development Ltd.).

From the complete matrix for incompressible flow (in the absence of thermo-dynamic influences) we thus obtain as dimensionless parameters those charac-teristic numbers which play an important part in scientific model testing. There are five of these, namely, the difference between eight physical quantities and three basic units: $8 - 3 = 5$. This relationship is no mere coincidence, but can be formulated as a law:

The number of dimensionless parameters contained in a complete set is equal to the number of physical quantities considered minus the rank of the matrix determined by the requisite basic units (Buckingham's theorem).

To give the proof of this law would be outside the scope of this treatment of the subject. However, there is an important point to be noted: the matrix presented in the first of the two above examples could alternatively, and quite suitably, have been established for the mass system. In that case, it would have been a matrix of the third rank. Obviously, the same number of dimensionless products could have been derived from it, i.e. the number would have remained unchanged, although according to Buckingham's theorem it ought to be reduced by one.

This apparent contradiction is due to the fact that with the mass system the number of basic units necessary for describing the physical phenomenon is superabundant, since the problem under consideration is purely a statistical one in which time is of no real significance. There is indeed no need to introduce Newton's law for formulating it, and time as a basic unit is redundant in this instance.

2.1.4 MODEL LAWS OF STATICS

The object of these considerations is to deduce the similitude or transfor-mation laws interlinking mechanical models of different dimensional scale and to understand these laws. Some additional quantities have to be intro-duced. The following is a list of the physical quantities employed, together with the symbols representing them:

(a) Quantities to be determined:	reaction	R	1
	stress	σ	2
	strain	ε	3
	displacement	δ	4
(b) Model characteristics:	length	l	5
	modulus of elasticity	E	6
	density	ρ	7
	Poisson's ratio	μ	8
	linear thermal expansion	α	9

(c) Loading actions:

	load	P	10
	pressure	p	11
	temperature	t	12
	gravitational acceleration	g	13
	load application co-ordinates	x, y, z	14, 15, 16

(d) Initial conditions:

	self-induced stress	σ_0	17
	forced deformation	u_0	18

All the physical quantities listed here can, in principle, be reproduced in the model to a scale different from that in the prototype. The similarity ratio linking model and prototype is given a distinctive subscript to denote the physical quantity to which it refers; the ratio for lengths (linear scale), which is of dominant importance for geometric similarity, forms an exception in that it has no subscript. Examples:

$$\text{linear scale} \quad L' = \lambda l$$
$$\text{scale of forces} \quad P' = \lambda_p p$$
$$\text{time scale} \quad T' = \lambda_t t$$

A simple example from structural analysis will help to explain the relationship between model mechanics and dimensional analysis (Fig. 2.3).

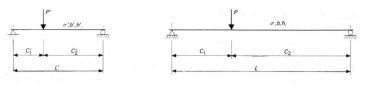

Figure 2.3

Consider two geometrically similar beams loaded at corresponding points. The stresses will be determined at corresponding points.

The stresses in the beams are respectively:

$$\sigma = \frac{M}{W} = \frac{Pc_1c_2}{l} \frac{6}{bh^2}; \qquad \sigma' = \frac{M'}{W'} = \frac{P'c_1'c_2'}{l'^2} \frac{6}{b'h'^2}.$$

Because of the presupposed geometric similarity, the lengths c, l, b and h for model and prototype can be transformed into one another by means of the similarity ratio λ. We can substitute

$$\frac{c_1c_2}{lbh^2} = \lambda^2 \frac{c_1'c_2'}{l'b'h'^2}$$

into the equations for σ, whence we obtain:

$$\sigma = \frac{Pc_1'c_2'}{l'b'h'^2}6\lambda^2 = \sigma'\frac{P}{P'}\lambda^2$$

or:

$$\lambda^2\frac{\lambda_\sigma}{\lambda_p} = 1.$$

In the relationship thus derived, all the specific properties associated with the example—such as the number and positions of the loads, and the cross-section at which the stress was calculated—have disappeared. All that remains is a relationship linking the physical quantities occurring in this example. These are, evidently, the same quantities as those which would be introduced in a dimensional analysis.

We shall now establish the dimensional matrix for this same problem. The quantities introduced and the desired quantities are confined to the geometric features embodied in l, b and h, the load P (whose point of application is determined, for example, by c) and the stress σ. The matrix thus becomes:

	l	P	σ	b	h	c_1
L	1	0	-2	1	1	1
F	0	1	1	0	0	0

The $6 - 2 = 4$ dimensionless parameters are:

$$\frac{\sigma l^2}{P}; \quad \frac{b}{l}; \quad \frac{h}{l}; \quad \frac{c_1}{l}.$$

What have we now gained by determining the dimensionless parameters of powers of the basic units?

The general solution of the problem posed in our example and aimed at interlinking the six given quantities can be represented as an unknown function:

$$f_1(l, P, \sigma, b, h, c_1) = 0. \tag{2.3}$$

Since this function must, in conforming with the laws of physics, be dimensionally homogeneous, it can always also be represented as a function of the dimensionless variables found by means of dimensional analysis, i.e. in the form:

$$f_2\left(\frac{\sigma l^2}{P}, \frac{b}{l}, \frac{h}{l}, \frac{c_1}{l}\right) = 0. \tag{2.4}$$

We have thus established the model law, for equation (2.4), which is equivalent to equation (2.3), additionally introduces the general condition for the validity of physical laws to which equation (2.3) would also have conformed. But equation (2.4) makes a more general statement in that it does not specify

any absolute quantities but instead requires only ratios. In other words, independent of size and loading, two structures whose behaviour is determined by the six physical quantities that we have introduced will behave similarly so long as the four ratios comprised in equation (2.4) are maintained.

On comparing the model with the prototype, we can write the conditions:

$$\frac{\sigma' l'^2}{P'} = \frac{\sigma l^2}{P}; \qquad \frac{b'}{l'} = \frac{b}{l}; \qquad \frac{h'}{l'} = \frac{h}{l}; \qquad \frac{c_1'}{l'} = \frac{c_1}{l}.$$

The last three conditions merely signify that there is geometric similarity of the model and that the point of application of the load is correspondingly located. The first condition expresses in a more general form what has already been found in our particular example. Equation (2.4) may be written in the form:

$$\sigma = \frac{P}{l^2} f_3 \left(\frac{c_1}{l}, \frac{h}{l}, \frac{b}{l} \right)$$

and thus in fact shows the general structure of all formulae for calculating the stresses in beams subjected to point loads. The range of validity of this simple transformation relationship, which as has been shown can quite conveniently be deduced from a particular example, is of course extremely limited. In cases where general problems in similarity mechanics are concerned, the transformation laws may be so unwieldy that any attempt to derive them more or less intuitively from a specific example is liable to be difficult and indeed hazardous. In this context we need only think of structures composed of a combination of several materials and consider also such complications as plate-buckling problems, creep, thermal stresses, etc. The overall problem becomes even more complex if flow problems and structural problems are subject to combined investigation in one and the same experiment, e.g. oscillation tests on suspension bridges, flutter of aircraft wings or suspended roofs, etc.

With dimensional analysis we can conveniently find out about the possible model laws. In such circumstances there is no disadvantage in introducing initially too many, rather than too few, physical constants into the dimensional matrix. When the relationships determined in this way are applied to a particular model problem, it will become apparent whether one or other similarity condition can be discarded.

We shall now establish a dimensional matrix possessing the widest possible applicability in the domain of model statics and dynamics and consider the similarity conditions that emerge from it. As already stated, we are free to introduce any number of quantities. However, for convenience, the physical quantities that we shall introduce will be confined to those mainly encountered in model-testing practice.

In order to obtain more direct information on the four quantities to be determined, we shall consider only the matrix comprising the model characteristics, the loads and the initial conditions, together with, in each instance, one of the

required quantities. We can thus write the following four equations:

$$\frac{R}{P} = g_1\left(\frac{EI^2}{P},\frac{\rho gl}{E},\mu,\alpha t,\frac{pl^2}{p},\frac{x}{l},\frac{y}{l},\frac{z}{l},\frac{\sigma_0}{E},\frac{u_0}{l}\right) \tag{2.5}$$

$$\frac{\sigma l^2}{P} = g_2\left(\frac{EI^2}{P},\frac{\rho gl}{E},\mu,\alpha t,\frac{pl^2}{P},\frac{x}{l},\frac{y}{l},\frac{z}{l},\frac{\sigma_0}{E},\frac{u_0}{l}\right) \tag{2.6}$$

$$\varepsilon = g_3\left(\frac{EI^2}{P},\frac{\rho gl,}{E},\mu,\alpha t,\frac{pl^2}{P},\frac{x,y}{l},\frac{z}{l},\frac{\sigma_0}{E},\frac{u_0}{l}\right) \tag{2.7}$$

$$\frac{\delta}{l} = g_4\left(\frac{EI^2}{P},\frac{\rho gl}{E},\mu,\alpha t,\frac{pl^2}{P},\frac{x}{l},\frac{y}{l},\frac{z}{l},\frac{\sigma_0}{E},\frac{u_0}{l}\right) \tag{2.8}$$

	R	σ	ε	δ	l	E	ρ	μ	α	P	p	t	g	x	y	z	σ_0	u_0
L	1	-1	0	1	1	-1	-3	0	0	1	-1	0	1	1	1	1	-1	1
M	1	1	0	0	0	1	1	0	0	1	1	0	0	0	0	0	1	0
T	-2	-2	0	0	0	-2	0	0	0	-2	-2	0	-2	0	0	0	-2	0
t	0	0	0	0	0	0	0	0	-1	0	0	1	0	0	0	0	0	0

1. $\dfrac{R}{P}$ 2. $\dfrac{\sigma l^2}{P}$ 3. ε 4. $\dfrac{\delta}{l}$ 5. $\dfrac{EI^2}{P}$ 6. $\dfrac{\rho gl}{E}$ 7. μ 8. αt

9. $\dfrac{pl^2}{P}$ 10. $\dfrac{x}{l}$ 11. $\dfrac{y}{l}$ 12. $\dfrac{z}{l}$ 13. $\dfrac{\sigma_0}{E}$ 14. $\dfrac{u_0}{l}$

Equations (2.5) and (2.6) for R and σ contain the external loading P in an explicit form. If reactions and stresses occur without the action of external loads (e.g. as a consequence of temperature effects or settlement), the quantity R/P must be replaced by R/EI^2 and the quantity $\sigma l^2/P$ by σ/E, which is permissible by combining with EI^2/P. Of course, alternatively, the desired quantities may be interchanged with other variables.

The dimensionless parameters will now be considered in somewhat closer detail and their significance in terms of model analysis will be discussed.

1. $$\frac{E'l'^2}{P'} = \frac{EI^2}{P} \quad \text{or} \quad \frac{P'}{P} = \frac{E'}{E}\left(\frac{l'}{l}\right)^2.$$

In principle, this condition can easily be satisfied by a correct choice of the loading on the model. However, for technical reasons in connection with obtaining reliable measured values, it is in practice often desirable to increase the model loading to something in excess of the ratio required by this similarity condition. This is permissible only if the law of superposition is valid for the problem in question. The condition for this will be formulated later on.

2. $$\frac{\rho'g'l'}{E'} = \frac{\rho gl}{E} \quad \text{or} \quad \frac{\rho'}{\rho} = \frac{g}{g'}\frac{l}{l'}\frac{E'}{F}.$$

For equal modulus of elasticity E and equal gravitational acceleration, there will be a similarity in behaviour between prototype and model only if the specific gravity is in inverse ratio to the scale. To satisfy this condition is very difficult in practice. For model experiments on dams, which are structures in which the state of stress under dead weight is of major importance, special materials with large ρ and small E have been developed which enable the condition to be satisfied (Section 3.1.2). Another possibility exists in varying the magnitude of g by centrifuging the model.

3.
$$\mu' = \mu.$$

The requirement that Poisson's ratio should be of equal magnitude in model and prototype is a serious practical and theoretical problem since μ is a natural material constant which is largely outside the experimenter's control. This problem will be considered on several occasions in this book (Section 3.1).

4.
$$\alpha't' = \alpha t.$$

Although the introduction of a particular required temperature distribution presents considerable technical difficulties, there are no theoretical difficulties involved in satisfying this condition.

5.
$$\frac{p'l'^2}{P'} = \frac{pl^2}{P}.$$

This requirement is easy to fulfil and is of course subject to the same conditions as those stated above for Condition 1. Alternatively, it can be put in the form $p'/E' = p/E$.

6.
$$\frac{x'}{l'}, \frac{y'}{l'}, \frac{z'}{l'} = \frac{x}{l}, \frac{y}{l}, \frac{z}{l}.$$

This condition formulates the geometric similarity and the requirement that all corresponding loads must be applied at homologous points.

7.
$$\frac{\sigma'_0}{E'} = \frac{\sigma_0}{E}.$$

This is identical with $\varepsilon'_0 = \varepsilon_0$ and is a consequence of Condition 1 or 5. It signifies that the strains in the model and in the prototype are of equal magnitude.

8.
$$\frac{\delta'_0}{l'} = \frac{\delta'}{l}.$$

This follows likewise from 1 or 5 and is obtained by integration from Condition 7. The deflections are geometrically similar.

The matrix which has been examined above and the similarity conditions deduced from it are not just valid for structures subject to small deformations

(which is a limitation of a large proportion of the theories in structural analysis). They are also valid for highly flexible structures such as ropes or cables possessing flexural stiffness, and for thin beams and columns whose deflections are too large to be linearly proportional to the magnitude of the loads. Even quite small deformations may significantly affect the external loads. This occurs, for example, in the case of a flexural member simultaneously subjected to lateral loading and external axial loads. Membrane effects in plates (Section 3.1) also belong to this group of phenomena. Although the stress–strain relation may well be linear, such structures are classed as 'non-linear'. On the other hand, there are a great many practically important 'linear' structures in which all the deformations are linear functions of the loading. The law of super-position is valid only for linear structures. If linearity of structural behaviour is presupposed, all the reactions, stresses, strains and displacements can per-missibly be taken as being proportional to the loading, i.e. to P. On considering equations (2.5), (2.6), (2.7) and (2.8) again, and bearing in mind that R/P already embodies the proportionality of the function R to P, we can dispense with the more general conditions El^2/P and ε in this case.

2.1.5 EXAMPLES

2.1.5.1 Elastic-model test for a bridge structure

This represents a typical test of the kind frequently performed in model-testing laboratories. It will be assumed that the shape of the deck slab or deck plate can be chosen with a considerable degree of freedom, that the bridge is a continuous structure on a fairly large number of supports and that its cross-sectional design features are too complex (as in a multi-cell box girder) for practicable analytical calculation even with the aid of a computer.

It will further be assumed that the model is similar to the prototype but that exaggeration of the strains in the model is permissible (Section 3.1).

The commonly encountered forms of loading will be considered, namely area loading p_F, strip loading p_S, concentrated load P and bending moment. In each case, the similarity relationship with regard to significant quantities such as stress σ, reaction R, total moment M, deck-slab moment m and de-flection δ will be established.

The dimensional matrix for the quantities under consideration is as follows:

| | Loads | | | | | Required quantities | | | | | |
	p_F	p_S	P	M	l	σ	R	M	m	$(E\delta)$	$(E\varepsilon)$
L	-2	-1	0	1	1	-2	0	1	0	-1	-2
F	1	1	1	1	0	1	1	1	1	1	1

Instead of the deflection δ and strain ε, the products $E\delta$ and $E\varepsilon$ are respectively introduced as physical quantities. This is done because 'strain exaggeration' is permitted. If δ were taken as the single deformation quantity, it would involve the limiting condition $\delta:\delta' = l:l'$.

In this example we shall not only establish an independent set of (in this case $10 - 2 = 8$) combinations but also their products with the object of obtaining for this type of model test a complete assembly of formulae for the possible similarity conditions. Hence each of the four types of loading envisaged will be linked to all the quantities which are required.

Thus arranged, the following dimensionless relations can be established:

$$\text{for } p_F: \quad \frac{p_F}{\sigma} \; ; \quad \frac{p_F l^2}{R} \; ; \quad \frac{p_F l^3}{M} \; ; \quad \frac{p_F l^2}{m} \; ; \quad \frac{p_F l}{E\delta} \; ; \quad \frac{p_F}{E\varepsilon}$$

$$\text{for } p_S: \quad \frac{p_S}{\sigma l} \; ; \quad \frac{p_S l}{R} \; ; \quad \frac{p_S l^2}{M} \; ; \quad \frac{p_S l}{m} \; ; \quad \frac{p_S}{E\delta} \; ; \quad \frac{p_S}{E\varepsilon l}$$

$$\text{for } P: \quad \frac{P}{\sigma l^2} \; ; \quad \frac{P}{R} \; ; \quad \frac{Pl}{M} \; ; \quad \frac{P}{m} \; ; \quad \frac{P}{E\delta l} \; ; \quad \frac{P}{E\varepsilon l^2}$$

$$\text{for } M: \quad \frac{M}{\sigma l^3} \; ; \quad \frac{M}{Rl} \; ; \quad \frac{M}{M} \; ; \quad \frac{M}{m \cdot l} \; ; \quad \frac{M}{E\delta l^2} \; ; \quad \frac{M}{E\varepsilon l^3} \, .$$

With the aid of these, the similitude laws for the required quantities can now be written down:

$$\sigma = \sigma' \frac{p_F}{p_{F'}} \qquad = \sigma' \frac{p_S}{p_{S'}} \lambda \qquad = \sigma' \frac{P}{P'} \lambda^2 \qquad = \sigma' \frac{M}{M'} \lambda^3$$

$$R = R' \frac{p_F}{p_{F'}} \frac{l}{\lambda^2} \qquad = R' \frac{p_S}{p_{S'}} \frac{l}{\lambda} \qquad = R' \frac{P}{P'} \qquad = R' \frac{M}{M'} \lambda$$

$$M = M' \frac{p_F}{p_{F'}} \frac{l}{\lambda^3} \qquad = M' \frac{p_S}{p_{S'}} \frac{l}{\lambda^2} = M' \frac{P}{P'} \frac{l}{\lambda} \qquad = M'$$

$$m = m' \frac{p}{p_{F'}} \frac{l}{\lambda^2} \qquad = m' \frac{p_S}{p_{S'}} \frac{l}{\lambda} \qquad = m' \frac{P}{P'} \qquad = m' \frac{M}{M'} \lambda$$

$$\delta = \delta' \frac{E'}{E} \frac{p_F}{p_{F'}} \frac{l}{\lambda} = \delta' \frac{E'}{E} \frac{p_S}{p_{S'}} \qquad = \delta' \frac{E'}{E} \frac{p_S}{P_{S'}} \qquad = \delta' \frac{E'}{E} \frac{M}{M'} \lambda^2$$

$$\varepsilon = \varepsilon' \frac{E'}{E} \frac{p_F}{p_{F'}} \qquad = \varepsilon' \frac{E'}{E} \frac{p_S}{p_{S'}} \lambda = \varepsilon' \frac{E'}{E} \frac{P}{P'} \lambda^2 = \varepsilon' \frac{E'}{E} \frac{M}{M'} \lambda^3 \, .$$

It is helpful to visualise special cases in which the similarity relationship between two comparable quantities becomes independent of the scale of the model. It appears that this is always so if the dimension of the loading is in agreement with the dimension of the denominator in the ratio (in the above assembly of dimensionless quantities). Thus, for equal distributed loads in

model and prototype, the stresses will be identical; for equal concentrated loads, the reactions, shear forces and slab moments will be identical; and for equal external moments, the internal moments will be identical (at corresponding points). Presupposing the same modulus of elasticity E for model and prototype, the strains (corresponding to the stresses) will be identical for equal distributed loading, and the deflections will be identical for equal strip loading.

2.1.5.2 Earthquake loading test on a dam

The object of the test is to determine the safety against failure of a dam subjected to earthquake action taking account of the dynamic pressure effect of the water on the dam, i.e. the inertia of the storage reservoir relative to the seismic oscillation of the soil.

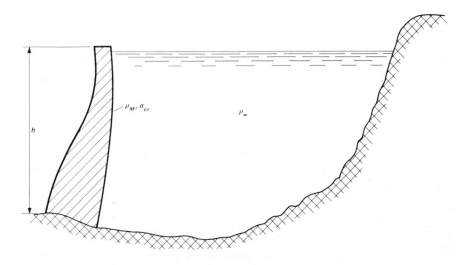

Figure 2.4

To start with it will be assumed that the reservoir and the dam are reproduced as a model, geometrically similar to the full-size prototype, on a vibrating table. By these means we may produce seismic oscillations homologous to those expected to occur in reality.

First, only the following physical quantities will be regarded as significant for the purpose of the model test; the height h of the dam as the geometric scale for the model, the amplitude a and the angular velocity ω for characterising the seismic oscillation, the density ρ_M of the dam and ρ_w of the water in the reservoir, and the failure stress σ_{cr} of the dam. Using these quantities, the

dimensional matrix may now be written (the significance of g and E will be discussed later):

	h	a	ρ_M	ρ_w	σ_{cr}	ω	g	E
M	0	0	1	1	1	0	0	1
L	1	1	-3	-3	-1	0	1	-1
T	0	0	0	0	-2	-1	-2	-2

Hence we obtain the three dimensionless quantities:

$$\pi_1 = \frac{\sigma_{cr}}{\omega^2 \rho_M a^2} \qquad \pi_2 = \frac{a}{h} \qquad \pi_3 = \frac{\rho_M}{\rho_w}.$$

If water is used as the liquid in the model reservoir, i.e. $\rho_w = \rho'_w$, it follows from the third condition

$$\rho'_M = \frac{\rho'_w}{\rho_w} \rho_M = \rho_M,$$

that the densities of the model dam and prototype dam materials must also correspond.

From the second similitude condition, we obtain for the ratio of the amplitudes:

$$a' = \frac{h'}{h} a = \lambda \cdot a.$$

This is a condition for controlling the motion of the vibrating table. If these conditions are satisfied, the following relation is obtained for the vibrating frequency of the model:

$$\omega' = \frac{\omega}{\lambda} \sqrt{\frac{\sigma'_{cr}}{\sigma_{cr}}}.$$

On the assumption that we have a 1/100 scale model, i.e. $\lambda = 1:100$, the model would have to be vibrated with one hundred times the frequency of the prototype if it were made of the same material ($\sigma_{cr} = \sigma'_{cr}$) as the latter while the amplitude would have to be one-hundredth of the actual amplitude of the seismic oscillations.

If the model analysis is limited to the physical quantities so far included in the similarity relationships considered here, it will not be difficult to see that the initial loading of the dam as a consequence of the static water pressure is not being taken into account correctly to scale. In the model this loading is exceedingly small (only 1% of that in the prototype), whereas in the prototype it contributes significantly to producing the fracture stress. As it is not permissible in a loading test to failure simply to add the seismic stresses to the known static stresses by superposition, a more discriminating model law must be established by the introduction of another physical quantity. By

appropriately controlling the gravitational acceleration g, it would of course theoretically be possible to alter the static pressure of the water on the model dam in the appropriate ratio.

On introducing the additional quantity g into the dimensional matrix, we obtain a further similitude condition:

$$\pi_4 = \frac{\sigma_{cr}}{g\sigma_M h} \quad \text{or} \quad \frac{h'g'\rho'_M}{hg\rho_M} = \frac{\sigma'_{cr}}{\sigma_{cr}}.$$

To achieve such a change of g would, for example, necessitate installing the whole model-testing arrangement in a centrifuge; since this is hardly practicable, we must assume that $g' = g$. This means, however, that we would have to make the model of a material whose failure stress would, for equal density, be only one-hundredth of that of the prototype. If no such material exists, the relation π_3 still offers the possibility of somewhat improving the failure to stress ratio by using a heavier liquid in the model:

$$\sigma'_{cr} = \lambda \sigma_{cr} \frac{\rho'_w}{\rho'}$$

though in this case the density of the material of the dam must be increased in the same ratio. The development of model materials having low strength in combination with high density is a problem that has received much attention from a number of laboratories concerned with the testing of structural models of dams.

So far, it has implicitly been assumed that the dam will behave like a rigid body and accurately follow the earthquake movements. For this reason,

Figure 2.5 Model of a dam (Diga Ambiesta) built on a vibrating table for testing. The reservoir is filled with water.

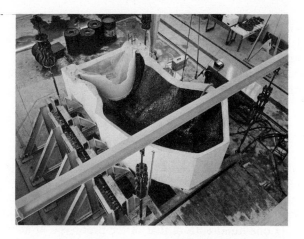

Figure 2.6   The model after destruction by simulated earthquake effect (I.S.M.E.S. Institute, Bergamo).

no deformation quantities have been introduced into the similarity relations. If the natural elastic (or characteristic) oscillation of the dam significantly influenced the failure behaviour, a further quantity, e.g. the modulus of elasticity, would have to be introduced into the dimensional matrix in order to characterise the elastic property. This in turn would impose another condition:

$$\frac{E'}{E} = \frac{\sigma'_{cr}}{\sigma_{cr}}$$

thereby making the choice or the development of a suitable model material even more difficult. However, in actual practice it is considered that because of the impact-like character of seismic loads ('earthquake shocks') this 'hydro-elastic' effect is of little significance and can therefore be neglected.

2.1.5.3   Aerodynamic testing

The vibrational behaviour of an elastic tower-type building is to be investigated with a model in a wind tunnel. The frequency and amplitude of the vibrations as well as the expected stresses are to be determined as functions of the approach velocity of the wind. The geometric features of the model are assumed to be similar to those of the prototype.

The following physical quantities, considered to be characteristic of the behaviour of the model, are introduced: the size l of the structure, the horizontal displacement δ, the strain ε, the stress σ, the mean density of the structural material ρ_T and of the air ρ_L, the modulus of elasticity E, the vibrational frequency ω and the wind velocity v.

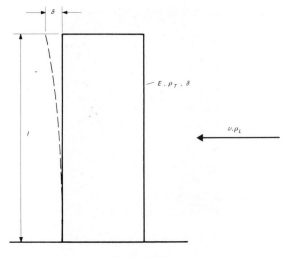

Figure 2.7

The dimensional matrix is:

	l	δ	ε	σ	ρ_T	ρ_L	E	ω	v
M	0	0	0	1	1	1	1	0	0
L	1	1	0	−1	−3	−3	−1	0	1
T	0	0	0	−2	0	0	−2	−1	−1

whence we obtain, for example, the following $9 - 3 = 6$ dimensionless relations:

$$\pi_1 = \frac{\delta}{L} \qquad \pi_2 = \varepsilon \qquad \pi_3 = \frac{\rho_T}{\rho_L} \qquad \pi_4 = \frac{\sigma}{E} \qquad \pi_5 = \frac{\omega l}{v} \qquad \pi_6 = \frac{\rho v^2}{E}$$

π_5 is the so-called Strouhal number which occurs in aerodynamics.

It will be assumed that the prototype is a concrete structure and that the model is made of, for example, a plastic. From π_6 we obtain for the corresponding wind velocity:

$$v' = v\sqrt{\frac{E'}{E}}$$

and in combination with the fifth condition this gives the corresponding vibrational frequency:

$$\omega' = \frac{\omega}{\lambda}\sqrt{\frac{E'}{E}}.$$

This is valid on condition that the third condition is also satisfied which, if the model is tested in a stream of air, presupposes an equal density for the model and prototype. Because the density of plastic is far lower than that of

Figure 2.8   Two aero-elastic test models.

Figure 2.9   The metal weights incorporated in the Plexiglas model to simulate the average density of the prototype (made of concrete) can clearly be seen in this photograph (National Physical Laboratory, Teddington).

concrete, the weight difference must be added into the model—uniformly distributed as much as possible (Fig. 2.6).

The deformations are true to scale and the stresses in the prototype are obtained from the second and fourth dimensionless relationships (since $\varepsilon' = \varepsilon$):

$$\sigma = \varepsilon'E.$$

It should be noted that in the similarity considerations presented here the Reynolds number has not been included. With aerodynamic models this is always permissible if the point of occurrence of separation of the flow, with consequent eddy formation, is pre-determined by the geometry of the specimen under investigation (e.g. sharp edges).

2.1.5.4 The water drop

In Section 2.1.6.3 the dependence of the shape of a body upon its absolute scale will be considered. This constitutes a departure from the principle of geometric affinity between the systems that are compared which is a requirement of similitude mechanics. The drop of liquid presents a striking example of the determining effect that size has upon shape and helps to clarify the concept of scale effect.

Consider a drop of water at rest on a dry base. Its shape is uniquely determined by its geometric size l, its density ρ, its surface tension σ and the gravitational acceleration (or more generally: the gravitational field quantity) g. From these data the shape of the drop can also be calculated analytically.

Figure 2.10

The dimensional matrix is as follows:

	l	σ	ρ	g
M	0	1	1	0
L	1	0	-3	1
T	0	-2	0	-2

and the dimensionless product

$$\pi = \frac{\sigma}{\rho g l^2}$$

is obtained, from which, for example, the following relation can be deduced:

$$l' = l\sqrt{\frac{\sigma'}{\sigma}\frac{\rho}{\rho'}\frac{g}{g'}}.$$

The surface tension and the density of water are physical material constants which cannot be varied within one and the same liquid. On the other hand, we know from the above relation that the (linear) dimensions of geometrically similar drops of the same liquid are inversely proportional to the square root of the acceleration due to gravity. Thus a drop of water on the Moon will be related to one on Earth if the former is 2.5 times as large as the latter. It is possible only within very narrow limits to make similar drops from different liquids since it is hardly practicable to alter the surface tension artificially in a suitably controlled manner. This fact clearly shows how the enforcement of geometric similarity for models can be contrary to Nature.

2.1.5.5 Speculative application of the similarity laws

So far we have applied the laws of similitude to mechnical systems that are relatively easy to visualise. In one sense the statement of the problem was always the same: to seek the conditions for a mechanical system to enable us to determine in a straightforward manner the behaviour of a geometrically similar system constructed to a different scale. Theoretically this should always be possible so long as we remain within the domain of physics, though in practice such attempts may fail because the similarity conditions are not realisable.

In Nature there are a great many phenomena which, while being essentially comparable, differ with regard to the scale on which they occur (Section 2.1.6). Although geometric similarity between these phenomena is hardly ever achieved exactly, it nevertheless appears tempting to apply considerations of dimensional analysis and model similarity to them. This leads to some surprising conclusions though admittedly these are inexact to the extent that the geometric similarity is only approximate. Yet they have a high content of probability and can thus be regarded as reasonable working hypotheses.

The following example will serve to illustrate the speculative application of dimensional analysis.

An elephant's heart performs 35 beats per minute, whereas the heart-beat frequency of a mouse is 500 per minute. Mammals of intermediate size are found to have frequencies intermediate between these two extremes. So it appears reasonable to suppose that there exists a relationship between the size of an animal and its heart-beat frequency and that this can be formulated as a law of similitude.

The mass m will be adopted as the most objective characteristic of the animal's size. The frequency is ω. The similitude law must be made dimensionless. We shall therefore establish the dimensional matrix and insert these two chosen quantities m and ω:

	m	ω	μ	g
M	1	0	1	0
L	0	0	-1	1
T	0	-1	-1	-2

Evidently is it not possible to form a dimensionless product with m and ω. Hence additional physical quantities must enter into the law of similitude that we are seeking to establish. What can these be? The function of the heart is to pump blood through the blood vessels. The flow of blood is laminar and it therefore appears reasonable to adopt as a further distinctive constant in considerations of similarity the viscosity μ of the blood. But with these three quantities it is still not possible to form a dimensionless parameter. By writing the ratio $m:\mu$ we do indeed get rid of the mass dimension and we are left with an expression which contains only length and time.

Could gravitational attraction, characterised by the field quantity g, exercise an influence on the rate of the animal's heart-beats? On introducing g into the matrix, we can now write down the following dimensionless parameter:

$$\pi = \frac{g\mu}{m\omega^3}$$

and thus arrive at the similitude law:

$$\omega' = \omega^3 \sqrt[3]{\frac{g'}{g}\frac{\mu'}{\mu}\frac{m}{m'}}.$$

Or, assuming $g' = g$ and $\mu' = \mu$:

$$\omega' = \omega\sqrt[3]{\frac{m}{m'}}.$$

It should be noted that the cube root in the ratio of mass to frequency has emerged as an unavoidable consequence of introducing g and μ. The surprising fact about this result is that the heart-beat frequency of mammals of different weight conforms almost exactly to this law deduced on the basis of abstract considerations. Is it therefore unreasonable to suppose that the rest of this law also possesses real validity? Omitting μ, this would signify that the heart-beat

Figure 2.11 This small monkey named 'Bonny' (weighing about 6 kg) was launched into earth orbit in a bio-satellite on 28 June 1969. The space flight was to have lasted 30 days, but was terminated after only nine days because the monkey's heart activity had become defective. It was brought safely back to earth, but died soon afterwards. Dr. W. Ross Adey, a UCLA physiologist, reported a progressive decline in the animal's blood pressure during the flight. This observation would appear to confirm the general validity of the similarity considerations presented in this chapter.

frequency is proportional to the gravitational acceleration. Actual observations have indeed confirmed that this conclusion, though not quantitatively exact, is at least qualitatively correct. The heart of the pilot of an aircraft pulling out from a fast dive is subject to a multiple of g and its beat rate increases tremendously. On the other hand, the heart activity of astronauts who spend a long time in a weightless condition diminishes quite appreciably.

It therefore appears to be a law of Nature that Man cannot live indefinitely in the absence of a gravitational field. Space research in the years ahead will probably confirm the truth of this speculative inference based purely on dimensional considerations.

2.1.6 SIMILARITY IN NATURE, TECHNOLOGY AND ARCHITECTURE

2.1.6.1 Observations

A familiar example often cited to provide the limitations of Man's technical ability to simulate natural structures is that of the cornstalk. The assertion that we are unable to construct a big tower having the slenderness of a cornstalk is indeed correct. Yet the comparison is misleading, for it ignores the fact that Nature itself would not be capable of constructing such a tower either. This paradox is not due to our technical impotence but to the similitude laws of mechanics.

If, with knowledge of the laws of model similitude, we observe Nature objectively, we shall frequently encounter these relationships between form and

scale provided that, by and large, we confine our considerations to geometrically 'similar' forms in Nature. We may, for instance, compare coniferous trees of various species with one another, or animals of the same general species. The last example in the previous section showed how similarity considerations can also be applied to living creatures. The size of the mammal was found to be related to its heart-beat frequency. A walk through a zoo will reveal how often the frequency (repetitive speed) with which certain periodic movements are performed is related to the size of the animal concerned. Here are some examples:

(a) Observing the wing-beat of the humming-bird, the crow and the eagle, it is evident that frequency decreases with increasing size of bird.

(b) A similar relationship is observed in the chewing movements of the mouse, the rabbit, the dog and the elephant.

(c) The sounds made by animals also provide an indication of size. Big animals roar or bellow, whereas small ones whistle or squeak. Here, too, we encounter an interdependence of frequency and geometric size.

A number of such examples could be given including many from within the sphere of our daily lives. The heart-beat of a young baby is nearly twice as fast as that of an adult. A small child suffers surprisingly little injury from a fall that would be dangerous to grown-up persons. This is due not so much to the intervention of some guardian angel as to the child's muscular cross-section in relation to its mass being substantially greater than that of the full-grown individual. Finally, consider a race between a five year old and a tall sprinter. The frequency of the child's leg movements is much faster than the adult athlete's yet the latter will win the race. Evidently high frequency is not necessarily equivalent to high speed of locomotion. The same conclusion can be drawn with regard to the wing-beat of the three types of bird referred to above. A humming-bird would certainly be the loser in a race against an eagle.

We shall now try to find the similitude law for this observation. To this end, let us, besides the frequency, again introduce the mass m as the reference quantity of comparison. We shall not specifically consider the heart function, so that the introduction of the viscosity (of the blood) for physical comparison is pointless. Instead we shall adopt as the overall quantity representing motive performance, the power N (rate of work done). In establishing the dimensional matrix, the geometric scale (the length L) must be introduced in order to arrive at a dimensionless expression. In conformity with the statement of the problem, we now seek a relation between this quantity and the speed.

From the dimensional matrix:

	m	N	L	ω	v
M	1	1	0	0	0
L	0	2	1	0	1
T	0	-3	0	-1	-1

we can, for instance, infer the existence of the two following dimensionless expressions:

$$\pi_1 \rightarrow \frac{N}{m} \frac{1}{l^3 r^3} \quad \text{and} \quad \pi_2 = \frac{vr}{l}.$$

The quotient N/m represents the specific power referred to the mass. It appears reasonable and permissible to assume this quotient to be at least approximately independent of the scale. On introducing this simplification, we obtain the following model laws:

For the frequency:
$$\omega' = \omega \sqrt[3]{\frac{1}{\lambda^2}}.$$

For the speed:
$$v' = v\sqrt{\lambda}.$$

These relationships do indeed state precisely what we have actually observed. They also present the correct order of magnitude in quantitative terms. To cover a distance for which the adult runner needs 10 seconds the boy half his size needs 14 seconds, and during this length of time the boy performs more than twice as many leg movements.

Generally speaking, biologists are unfamiliar with similitude mechanics. Yet the systematic application of these principles to the energy balance of living organisms would undoubtedly lead to some striking conclusions and add to our knowledge of natural phenomena. It must once again be emphasised, however, that all transformation laws can have proper validity only if there exists geometric similarity between the mechanisms. In Nature, this condition is never accurately satisfied. On the contrary, as we shall see later, the requirement for affinity between the outward forms of mechanisms or creatures differing in scale is, if anything, contrary to Nature. Nevertheless, similarity considerations can yield useful information if their field of application is selected with proper care.

If such considerations can be applied to biology, a field in which shape and form are only rarely conditioned by mechanical conditions, the laws of similitude must surely have a wide field of application in technology. Here are some technological phenomena which present a parallel with the biological examples mentioned above:

(i) The rotational speed of the engine of a small helicopter is higher than that of a large one, yet the larger machine generally attains higher flying speeds.

(ii) The smaller an internal combustion engine, the higher is its rotational speed. Compare the insect-like buzzing of a model aircraft engine with the steady pounding of a large marine diesel. The same general principle appears to apply to all prime movers including electric motors.

(iii) A very striking example of the relationship between the frequency of repetitive processes and the size of the machine is the development of the electronic computer. The tremendous increase in processing speed is a direct consequence of the miniaturisation of the electrical and electronic components used in computers.

Here again the number of examples could be increased almost indefinitely.

2.1.6.2 Geometric similarity is contrary to nature

Model analysis (or, more generally, model mechanics) predicts the behaviour of a mechnical system by studying a 'true' model made to a different (usually a smaller) scale from that of the prototype. We adopt this procedure because a model is much cheaper to construct than a full-size prototype, and also because the size of the model can be chosen to suit our laboratory equipment so that permanent testing facilities can be applied to a wide range of structures. As already stated, the model must—apart from a few exceptions that are unimportant in the context of these considerations—be geometrically similar to the actual system.

In a geometrically similar model system, it often proves very difficult to satisfy conditions imposed for the other physical quantities by the laws of similitude. This fact suggests that a given system, composed of a particular material and having specific geometric proportions, must occupy a natural place within the range of sizes or orders of magnitude. In a sense, by performing a model test we are (though quite deliberately and with a proper knowledge of the laws concerned) violating Nature. This emerged quite clearly in the example of the earthquake test on a dam (Section 2.1.5.2), in which artificially contrived combinations of materials had to be employed in order to satisfy the similarity conditions. It also emerged in the example of the water drop, in which we were unable adequately to influence the nature of the surface tension (Section 2.1.5.4). These difficulties manifest themselves in the 'scale effects' (Section 2.1.1). The lack of realism of similar scaling-down can be observed even in simple examples of model statics. When investigating the load capacity of a reduced-scale model made of the same material as the prototype, we must apply additional external loading to the model in order to reproduce equal dead load stresses. In other words, the model has a significantly better self-supporting capacity than the prototype. Alternatively, if the model were not simply a means to an end but had to perform a structural function in its own right, it could be made substantially more slender. Having established this point, in order to gain a better understanding of natural phenomena with the aid of general similitude laws, we can now pose the question the other way round:

For theoretical reasons, experimental model mechanics must fulfil the requirement of geometric similarity. It is thus obliged to make the other relationships demanded by the laws of similitude conform as far as possible. Nature, on the other hand, being concerned only with prototypes, is free to ignore this geometric condition and seeks to achieve the shapes of its structures by using natural means.

2.1.6.3 Shape related to dimensional scale

The following considerations do not strictly conform to the stated requirements of scientific model mechanics. Yet they are useful to the scientist or engineer concerned with the precise concept of the 'similar model', in that

they help him to realise how contrary to Nature (viewed over a wide spectrum of size) the geometric affinity of external forms differing in scale really is.

A simple example can help us to visualise this. Consider a concrete cube of edge length a placed on a short concrete column of square cross-section with sides of length d. The cube and the column are made of the same concrete, characterised by the permissible stress σ_p and the specific gravity s. Instead of keeping the ratio d/a constant when comparing cube-and-column assemblies of different sizes, as was done in previous examples of geometric similarity, we now wish to determine the absolute magnitude of d in relation to the size of the cube for constant stress in the column.

If we neglect the self-weight of the column, the stress in it is given by the relation:

$$\sigma = \frac{G}{F} = \frac{sa^3}{d^2}.$$

Putting $\sigma = \sigma_p$ and calculating the ratio d/a as a function of the edge length of the cube, we find:

$$\frac{d}{a} = \sqrt{\frac{s}{\sigma_p}}\sqrt{a} = k_1\sqrt{a}.$$

For any particular choice of material the first square-root value is a constant k_1, while $\sqrt{a}$ fixes the ideal value of d/a (and therefore the shape criterion) as a function of the absolute size of the cube. On substituting real values into the formula and adopting $s = 2.5t/m^3$ and $\sigma_p = 2500t/m^2$, we obtain $k_1 = \frac{1}{32}$, resulting in the following proportions for our idealised assembly (Fig. 2.12).

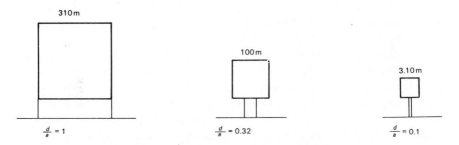

Figure 2.12   Cubes.

We have thus found a general law of Nature for the shapes of structures: the smaller the supported load, the more slender—i.e. the relatively finer or thinner—the supporting structure.

This is the fundamental reason why a fly can walk on a ceiling, whereas an elephant, even with the most powerful suction pads on its feet, could never perform such a stunt. Figures 2.13 and 2.14 provide a striking illustration of this general law of Nature. The legs of the elephant are not only thicker than those of the grasshopper in absolute terms but also relative to the bulk of the animal's body. Yet relatively speaking the grasshopper's hind legs perform a much more spectacular function. The spider's legs can be even much more

slender. If we envisage a grasshopper as large as an elephant, we know intuitively that legs of such slenderness, even when scaled up to elephant size, could not support the huge body. And if we imagine an animal larger than the elephant, able to walk about on land, we can readily conceive that its legs (the load-bearing elements) would become increasingly dominant in size relative to the bulk of the creature's body and that, above a certain size, such legs would no longer be able to support even their own weight. It has been suggested that this was the reason why the dinosaurs became extinct. The absolute order of magnitude—more particularly the body size—of living creatures is no mere coincidence: it is, instead, determined by the sort of

Figure 2.13 A comparison of this photograph of a grasshopper with that of an elephant (below) strikingly shows how impossible the bodily proportions of insects would be in the mammal-size animal world.

Figure 2.14 Elephants (Zoological Gardens of Basle).

construction materials available to Nature and by the gravitational force of the Earth.

In the above example of the cube on the column we can substitute for the weight the quantity $G = mg$ (gravitational force = mass × gravitational acceleration) while we can introduce the density ρ in lieu of the specific gravity s. We thus obtain the following equivalent expression for the ratio:

$$\frac{d}{a} = \sqrt{\frac{\rho}{\sigma_p}} \sqrt{a \cdot g}.$$

We thus see that in Nature optimal geometric proportions increase with the square root of the gravitational acceleration. If the Moon were inhabited by living creatures, we could expect them, on average, to be about 2.5 times as large as those found on Earth. If there were man-shaped creatures living on the planet Jupiter ($g = 27\,\mathrm{m\,s^{-2}}$), they would be about 80 cm high. If they were truly geometrically similar to Earth-men, i.e. scaled-down models of ourselves, they could move much more quickly, think more rapidly and have shorter lives. A bridge built on the Moon from the same materials as might be used on Earth would be correctly proportioned if all its dimensions were increased 2.5 times in comparison with such a bridge built on Earth.

Let us, finally, consider once more our example of the cube on the column. We see that for any particular material there is a limiting size above which this structure ceases to be self-supporting, i.e. it can no longer carry its own weight. To build a higher structure we are obliged to alter its shape. For example, instead of the cube we can perch a pyramid three times the height of the cube, and having the same square plan area as the latter, upon the supporting column, which will be stressed only to the permissible value. This compulsory choice of upward-tapering shapes for static structures which have to support only their own weight is suggestive of the geological formation of mountains. Generally speaking, higher mountain ranges have sharper peaks. This principle of a limiting size may be demonstrated quantitatively. Overhanging masses of rock are mostly of the order of magnitude of some tens of metres; vertical rock walls seldom exceed a few hundred metres in height; pyramid-shaped mountains rise thousands of metres into the air. The Earth and all other known celestial bodies are essentially spherical in shape. This is the only statically possible shape that enables such huge masses of matter to hold together. Any other shape would break up under the action of gravity. On the other hand, small meteors and meteorites, being much smaller, are not restricted to a spherical shape. On consideration, the whole range of shapes observed from the bizarre structures revealed by the electron microscope to the simple basic spherical shape of our own planet unmistakably demonstrate the existence of a definite interdependence between the shapes of possible structures and their absolute sizes.

Through the ages architecture has concerned itself with the problem of proportion, of correctness of 'scale'. The higher civilisations evolved their laws of proportion, ranging from the ancient Egyptians, Phoenicians and Aztecs

to the 'golden section' of the Renaissance and, in our own century, Le Corbusier's 'Modulor'. For a proper critical assessment of these 'rules', it must be realised that the principal dimensions of the relevant structures represent only a small section of the whole range of scale accessible to us (Fig. 2.15). From this diagram it can also be seen that models made for the investigation of structures do not, in terms of the scale range, differ greatly from their prototypes. Indeed, it is this comparative nearness in scale that makes the investigation of geometrically similar models meaningful.

The rules of architectural proportion, both ancient and modern, have one basic feature in common. They are purely geometric principles of co-ordination which aim to bring harmony to architectural design by reference to some 'absolute' standard dimension regarded as eternal and immutable—derived from astronomy or astrology or, more obviously, from the dimensions of the human body itself.

On reading how Le Corbusier evolved his concept of the Modulor, we can follow step by step the great architect's quest to establish the ultimate true scale for his structures and its application. Influenced by the principles of harmony in music, he searched for true visual harmony in the dimensional sub-division of space, with the very practical ulterior aim of achieving standardisation of architectural dimensional units. Other than the fact that he took the human body as his starting point, his procedure was a very abstract one taking no account of the real materials with which the shapes must ultimately be built. However, if it is the architect's main purpose to produce a conceptual space primarily suited to human needs and functions, the application of such rules can be entirely meaningful. To facilitate industrialised production by standardisation can also be a commendable aim.

But the application of such 'rules of harmony' in architecture may become hazardous if the designer attributes to them a truth content, a claim to absolute rightness, transcending their basis (e.g. human proportions) and verging on the mystic. In practice, such rules are merely convenient and useful conventions.

In this section, we have shown that geometric similarity rules are contrary to Nature. The material world is governed by more highly differentiated laws than spatial ones. Rules of proportions should be applied only to the order of scale for which they were established and must on no account be arbitrarily extrapolated.

Just as Nature may choose not to make full use of the full potential of structural elaboration from a given construction material and scale but instead select a simpler shape (for instance, a slug rather than a grasshopper), so the architect must be free to base his design on other than physical considerations (such as the above-mentioned rules), provided that he does not exceed the limit of what is materially possible.

On the other hand, the engineer will generally design structural forms close to the limiting boundaries of what is physically possible. For him there are no geometric rules *a priori*: geometric form is for him a consequence of his understanding of the strength of the structural material and the scale of the job.

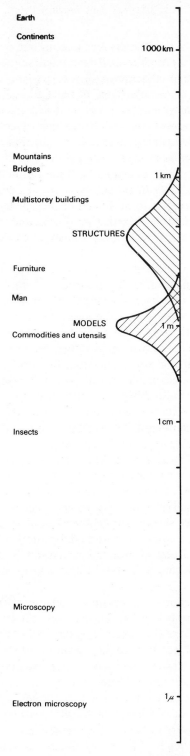

Figure 2.15 Absolute orders of magnitude.

Wachsmann, a disciple of Le Corbusier, produced designs for large structures evolved from purely geometric considerations. They were never built. Mies van der Rohe reached (and overstepped) the boundaries of what most engineers would regard as the architect's province: irrespective of scale, he applied the same proportions for spatial configuration for the load-bearing structure to anything from a small private house to a large exhibition hall, thus disregarding the natural laws of shape. In a sense, he constructs models having the reverse algebraic sign and accordingly has to contend with scale effects.

As reasoned many times in this book, the optimum sizes of structures designed on purely statical (and therefore also largely economical) principles depend on material and scale. Here lies the main drawback of mass prefabrication of long-span structures. It is also the reason why individual calculations and designs are made for most of the many bridges built every year.

2.2 STRESS AND STRAIN

2.2.1 SIGNIFICANCE OF DEFORMATION MEASUREMENTS ON MODELS

The main pattern of practical structural analysis is primarily aimed at establishing by direct means (if possible without a detour via deformation calculations) the equilibrium of external loads with internal stresses. Determination of the actions to which the material is subjected is extremely important. The determination of deformation quantities is usually regarded as merely an intermediate stage in the analysis of statically indeterminate systems and in solving differential equations arising in connection with elastic theory. Only in exceptional cases is the designer interested in absolute values of the deformations.

Model analysis, too, has the aim of finding the internal stresses occurring in the construction material. But by the very nature of the available measuring methods these stresses can only be determined indirectly, via deformation measurements. Strictly speaking, this applies even to apparently direct stress-analysis techniques such as photo-elasticity.

Thus, for the correct interpretation of the results of measurements obtained on elastic models, it is even more important than in ordinary structural analysis to have a close understanding of the stress–strain relationships. For the purpose of the following treatment of the subject, the reader will be assumed to be familiar with a number of standard problems of elastic theory.

The choice of the measuring positions on the model and the arrangement of the transducers (strain gauges, etc.) are largely left to the experimenter's discretion and judgment. Whether he will, with the arrangement chosen, be able to obtain the desired information concerning the structural behaviour of the specimen under investigation will depend on his skill and experience. The prior condition for the efficient arrangement of the measuring transducers is that the experimenter must be able to visualise in advance the sort of stress patterns that are fundamentally likely to occur in those parts of the structure in which he is particularly interested.

It is not within the scope of this book to explain the fundamentals of elastic theory. However, because of the special approach to the subject that the nature

of model analysis involves, the principal definitions and analytical relationships of elasticity will be repeated here in a somewhat unconventional manner
suited to the requirements of model testing. At the outset, it should be noted
that a distinction should be made between two types or stages of hypotheses
in the elastic theory: (a) the generally valid idealising assumptions as to
material properties, which are summarised in Section 2.2.2, and (b) the special
additional hypotheses which are, in various ways, adopted as a basis for the
solution of sub-problems arising in elastic theory to as to make them amenable
to mathematical treatment (Section 2.3).

2.2.2   THE DEFINITIONS OF E AND μ

All the elastic-theory considerations presented in this book are based on the
following idealising conceptions concerning the material:
   (i) The material is a homogeneous, isotropic continuum, i.e. its properties
are invariably the same, even in a small element selected from this material,
irrespective of orientation.
   (ii) The material shows completely elastic behaviour, i.e. any deformation
produced by the action of external forces will disappear completely on removal
of those forces.
   (iii) The material conforms to Hooke's law.
Consider a cylinder, with elastic properties as outlined above, which is
subjected to a direct tensile test (Fig. 2.16). Under the action of the force P,
which is visualised as a uniform stress σ_1 acting upon the end faces of the cylinder,
the latter undergoes an extension Δl and reduction in radius of Δr. Let ε_1 denote
the strain (relative extension) in the longitudinal direction, i.e. the direction
in which the stress is acting, and let ε_2 denote the transverse (or lateral) strain
perpendicular to the direction of the stress: $\varepsilon_2 = \Delta r/r$. The ratio of stress to
strain, produced by that stress in its direction of action, is called the *modulus
of elasticity* E; thus: $E = \sigma_1/\varepsilon_1$. It is a direct measure of the stiffness of the
material.
   The transverse strain, observed in nearly all materials and occurring without
the action of stress in the transverse direction, will in our material be assumed
to be a fixed ratio

$$\frac{\Delta r/r}{\Delta l/l} = -\frac{\varepsilon_2}{\varepsilon_1} = \mu$$

to the longitudinal strain and be invariant, irrespective of the magnitude of
the strains produced. This ratio μ is known as Poisson's ratio. Its magnitude
is a measure of the degree to which the material under consideration maintains
its volume when it has a deformation imposed upon it.

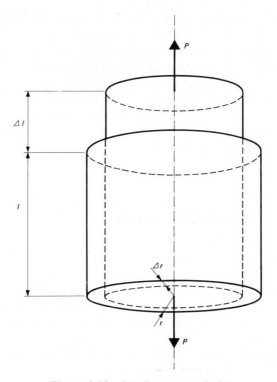

Figure 2.16 Tensile test on cylinder.

The relative volume change of our cylinder as a consequence of elastic deformation is:

$$\frac{\Delta V}{\pi r^2 l} = \frac{\Delta l}{l} + \frac{2\Delta r}{r} = \varepsilon_1 + 2\varepsilon_2 = \varepsilon_1(1 - 2\mu).$$

When Poisson's ratio $\mu = 0.5$, the volume of the specimen remains constant. This result is more particularly applicable to incompressible fluids whose modulus of elasticity for unrestrained transverse strain, as defined, is zero. Since the volume of an elastic specimen in a tensile test can obviously never be reduced, the values of μ for all materials must lie between 0 and 0.5.

These are the concepts by which elastic theory normally operates. In exceptional cases, one or other of the idealised notions (e.g. the assumption of hysteresis or of anisotropy) is modified for the investigation of special problems.

One of the main functions of elastic-model technology is in finding natural materials, or developing new artificial ones, whose behaviour closely conforms with the basic concepts defined above. Strict conformity can never be achieved because the microstructure of every material is characterised by a heterogeneous anisotropic conglomeration of individual particles (aggregates, crystals, molecules), so that the concept of a continuum is applicable only in a statistical sense (Section 3.1).

2.2.3 PLANE STRESS FIELD

For technical reasons, the state of deformation in model analysis is usually measured only at the surface of the test specimens. Here there are no stresses acting perpendicular to the surface except at the points of application of the forces so that on the exterior surface of the model there normally exists a plane state of stress. It is then usually possible, assuming such a state of stress at the surface, to infer the stress distribution inside the specimen from theoretical considerations (Section 2.3). To begin with, we shall confine discussion to the plane state of stress and deformation. We shall place a Cartesian co-ordinate system with its origin 0 at a point within the plane stress field and orientate its axes 1 and 2 in the directions of the principal stresses (Fig. 2.17). Next, we shall consider how a point P located an infinitesimal distance from 0 will move relative to this co-ordinate system as a result of distortion of the strain field. The position of P is determined by the vector r and the angle φ. The region in the vicinity of 0 is shown in Figure 2.18.

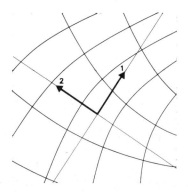

Figure 2.17

The state of strain around 0 is determined by the principal stresses, which are assumed to be known, and can be described by means of the principal strains:

$$\varepsilon_1 = \frac{\Delta x}{x} = \frac{1}{E}(\sigma_1 - \mu\sigma_2);$$

$$\varepsilon_2 = \frac{\Delta y}{y} = \frac{1}{E}(\sigma_2 - \mu\sigma_1). \tag{2.9}$$

When interpreting strain measurements, it is important to remember that, perpendicular to the *plane* stress field, a strain $\varepsilon_3 = -\mu(\sigma_1 + \sigma_2)$ must also occur. If this strain is prevented from developing, no plane state of stress is possible. Also, on transition from a three-dimensional to a plane state of stress

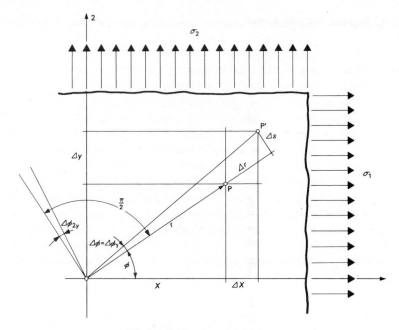

Figure 2.18   Plane stress field.

(e.g. at points of restraint), states of strain which are difficult to visualise are liable to arise which can sometimes make a simple interpretation of strain measurements impossible. Simplified assumptions as to the strain distribution inside the test specimen (e.g. Navier's hypothesis) are not permissible in such zones (Section 2.3).

We shall now describe the motion of P as a function of the angle φ. The relative position of the displaced point P' with respect to its original position P in the stress-free field is expressed by:

$$\Delta r = \Delta x \cos \varphi + \Delta y \sin \varphi\,;$$

$$\Delta s = -\Delta x \sin \varphi + \Delta y \cos \varphi. \tag{2.10}$$

We shall consider here only the relative movement in the direction of the radius vector; by substitution of equation (2.9) into equation (2.10) we obtain:

$$\varepsilon_\varphi = \frac{\Delta r}{r} = \frac{1}{E}[(\sigma_1 - \mu\sigma_2)\cos^2 \varphi + (\sigma_2 - \mu\sigma_1)\sin^2 \varphi]$$

$$= \frac{1}{E}[\sigma_1(\cos^2 \varphi - \mu\sin^2 \varphi) + \sigma_2(\sin^2 \varphi - \mu\cos^2 \varphi)] \tag{2.11}$$

or:

$$\varepsilon_\varphi = \varepsilon_1 \cos^2 \varphi + \varepsilon_2 \sin^2 \varphi. \tag{2.12}$$

We see that if the direction and magnitude of the principal stresses are known the distribution of the radial strain around a point can be described by a simple trigonometric relation. From equation (2.12) we can at once deduce another well-known relationship; since:

$$\varepsilon_{\varphi + \pi/2} = \varepsilon_1 \sin^2 \varphi + \varepsilon_2 \cos^2 \varphi$$

we obtain:

$$\varepsilon_\varphi + \varepsilon_{\varphi + \pi/2} = \varepsilon_1 + \varepsilon_2 \tag{2.13}$$

i.e. *the sum of two mutually perpendicular strains is invariant.*

In order to describe the displacement of the point P, we have considered only its movement in the radial direction and ignored its relative rotation in terms of the angle φ. We shall now demonstrate that the state of strain is completely described by the function (2.12) alone, which therefore already takes account of the rotation. We can write down the following expression for the rotation:

$$\Delta\varphi = \frac{\Delta s}{r} = -\frac{1}{E}(\sigma_1 - \mu\sigma_2)\sin\varphi\cos\varphi + \frac{1}{E}(\sigma_2 - \mu\sigma_1)\sin\varphi\cos\varphi. \tag{2.14}$$

Differentiation of equation (2.11) with respect to φ gives:

$$\frac{d\varepsilon_\varphi}{d\varphi} = \frac{1}{E}[-(\sigma_1 - \mu\sigma_2)\,2\sin\varphi\cos\varphi + (\sigma_2 - \mu\sigma_1)\,2\sin\varphi\cos\varphi]$$

$$= -\frac{2(1 + \mu)}{E}\sin\varphi\cos\varphi(\sigma_1 - \sigma_2). \tag{2.15}$$

On comparing equations (2.14) and (2.15) we can at once deduce:

$$2\frac{\Delta s}{r} = \frac{d\varepsilon_\varphi}{d\varphi}. \tag{2.16}$$

which proves that the movement of the point P is determined solely by the function $\varepsilon_\varphi(\varphi)$.

We shall now also consider the physical meaning of $\Delta s/r$. An inspection of Figure 2.18 shows that $\Delta\varphi_1$ and $\Delta\varphi_2$ evidently determine the shear strain $\gamma = \Delta\varphi_1 - \Delta\varphi_2$. It can be calculated from equation (2.14):

$$\Delta\varphi = \left.\frac{\Delta s}{r}\right|_\varphi = -\frac{(1 + \mu)}{2E}(\sigma_1 - \sigma_2)\sin 2\varphi$$

$$\left.\frac{\Delta s}{r}\right|_{\varphi - \pi/2} = \frac{(1 + \mu)}{2E}(\sigma_1 - \sigma_2)\sin 2\varphi = -\left.\frac{\Delta s}{r}\right|_\varphi$$

whence we obtain:

$$+\Delta\varphi|_\varphi - \Delta\varphi|_{\varphi - \pi/2} = -\frac{(1 + \mu)}{E}(\sigma_1 - \sigma_2)\sin 2\varphi = 2\,\Delta\varphi = -\gamma.$$

In retrospect, we thus see that the shear strain can be determined simply by differentiation of equation (2.16) with respect to φ:

$$\gamma = -\frac{d\varepsilon_\varphi}{d\varphi}.$$

With the definition equation for shear modulus—namely, $T = G\gamma$—it should also now be possible to find the relation between G and E.

The shear stress and the direction are linked by the well-known relation:

$$\tau = \tfrac{1}{2}(\sigma_1 - \sigma_2)\sin 2\varphi = G \cdot \gamma$$

and:

$$\frac{E}{2(1 + \mu)} = G.$$

It has thus been demonstrated in an unconventional manner that the state of strain can be fully and uniquely described in terms of the radial strain around a point, i.e. the shear modulus and the shear strain are merely additional concepts which are not essential to the definition of elastic theory. The shear stress is an auxiliary quantity whose only purpose is to describe the equilibrium in an element which is awkwardly orientated (i.e. deviating from the direction of principal stress).

The knowledge that ε_φ is sufficient to describe the plane stress and strain field is of fundamental importance with regard to measuring techniques since there are no instruments capable of giving direct angular measurements, i.e. without the detour via measurement of lengths.

The representation of strain in polar co-ordinates, as has been chosen here, has the advantage of making direct visual observation possible. This is of great importance in model analysis. The whole field of strain around a point can be surveyed at a glance. In planning a model test the arrangement of the measuring transducers must be chosen as efficiently as possible, and for this reason it is essential to obtain an insight into the strain conditions expected to occur. Information on the various states of stress and strain can be found from the relationships (2.11) and (2.12).

In Figures 2.19 and 2.20 and in Figures 2.21 and 2.22 a number of special cases of strain distribution with their associated stress patterns are shown.

The strain distribution diagrams have been drawn in accordance with the following convenient rules:

A circle around the point under consideration serves to establish zero strain, thus forming a reference line. The strains themselves are plotted radially with reference to this circle: positive on the outside (elongation), negative on the inside (shortening). A unit of strain corresponds to half the radius of the circle.

For the purpose of the diagrams the modulus of elasticity is assumed to have the magnitude of one stress unit, so that only the dimensionless effect of the angle and of Poisson's ratio is shown.

Figures 2.19 and 2.20 illustrate the distribution of strain ε_φ associated with the unit states of stress shown on the left, for various values of μ. The positions

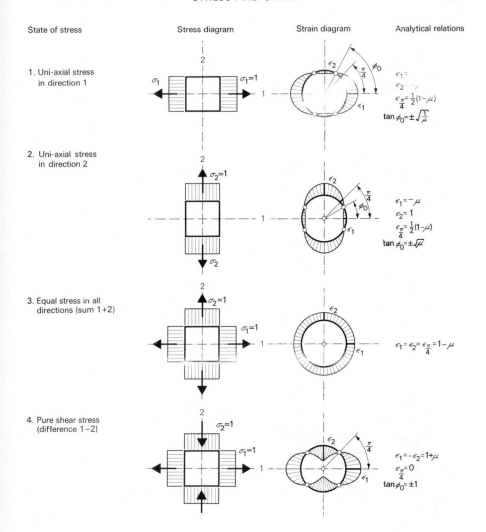

Figure 2.19 Strain distribution ε_φ for various states of stress (for $\sigma = 1$ and $\varepsilon = 1$; actual strains: $\varepsilon_{\text{eff}} = (\sigma_{\text{eff}}/E)\varepsilon$).

of the points of zero strain and the strain values at the angle $\pi/2$ are also indicated. If the diagrams are used for calculating the actual strain ε_{eff}, we must apply the relation:

$$\varepsilon_{\text{eff}} = \frac{\sigma_{\text{eff}}}{E}.$$

Figures 2.21 and 2.22 illustrate the stresses associated with the unit states of strain shown on the left, again for various values of μ. The effect of this latter quantity is clearly shown. The actual stresses σ_{eff} are obtained from:

$$\sigma_{\text{eff}} = E\varepsilon_{\text{eff}}.$$

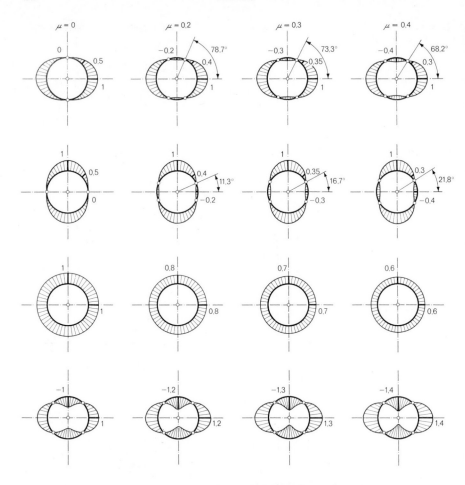

Figure 2.20   Numerical examples for various values of μ.

The diagrams may be superimposed upon each another in any desired combination so that all stress and strain states are determinable in this way. This possibility will be used repeatedly (Section 2.3).

From the strain distribution diagrams, it emerges that strain-free directions exist in conjunction with all sorts of stress distributions. It is helpful to be able to predict such directions in connection with the planning of a model test: they should, of course, be avoided as directions in which to perform measurements.

The condition for the existence of a strain-free direction will now be derived From equation (2.11):

$$\varepsilon_\varphi = \frac{1}{E}[(\sigma_1 - \mu\sigma_2)\cos^2\varphi + (\sigma_2 - \mu\sigma_1)\sin^2\varphi] = 0$$

State of strain	Strain diagram	Stress diagram	Analytical relations

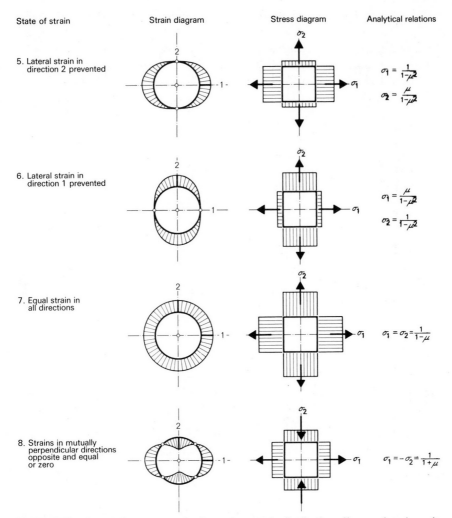

Figure 2.21   States of stress σ_1 and σ_2 for various strain distributions (for $\sigma = 1$ and $\varepsilon = 1$; actual stresses: $\sigma_{eff} = E\varepsilon_{eff}$).

we determine for the magnitude of the angle corresponding to zero strain:

$$\tan \varphi = \pm \sqrt{\frac{\mu - \dfrac{\sigma_1}{\sigma_2}}{1 - \mu \dfrac{\sigma_1}{\sigma_2}}}.$$

A real solution exists only if the following condition is satisfied:

$$\tan^2 \varphi \geqslant 0,$$

therefore, if $\mu - \sigma_1/\sigma_2 \geqslant 0$, then $1 - \mu\sigma_1/\sigma_2 \geqslant 0$ must also be true. From

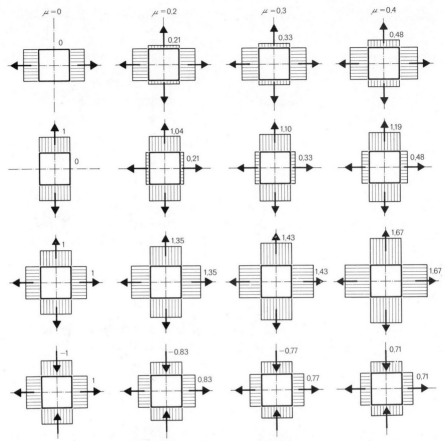

Figure 2.22 Numerical examples for various values of μ.

this we obtain as the first condition:

$$\mu \geqslant \frac{\sigma_1}{\sigma_2} \leqslant \frac{1}{\mu}$$

or, if $\mu - \sigma_1/\sigma_2 \leqslant 0$, then $1 - \mu\sigma_1/\sigma_2 \leqslant 0$ must also be true, whence the second condition is obtained:

$$\mu \leqslant \frac{\sigma_1}{\sigma_2} \geqslant 1/\mu.$$

If one of these conditions is satisfied, it means that a direction exists in which the strain becomes zero.

The foregoing considerations were based on a Cartesian system of coordinate axes disposed in the directions of the principal stresses. The existence of these preferential directions has tacitly been presupposed. For the sake of completeness, we shall justify the assumption that in the plane stress field there always exist two mutually perpendicular shear-free directions.

For this purpose it will be assumed that, in addition to normal stresses, tangential stresses (the shear stresses at the edges x and y) are also acting upon

the element represented in Figure 2.18. Examination of the equilibrium condition yields the well-known transformation equations:

$$\sigma = \frac{\sigma_x + \sigma_y}{2} + \frac{\sigma_x - \sigma_y}{2} \cos 2\varphi - \tau_{xy} \sin 2\varphi$$

$$\tau = \frac{\sigma_x - \sigma_y}{2} \sin 2\varphi + \tau_{xy} \cos 2\varphi.$$

Putting $\tau = 0$, in accordance with the assumption made hitherto, we obtain the condition:

$$\tan 2\varphi = \frac{2\tau_{xy}}{\sigma_y - \sigma_x}.$$

In the interval $0 \leqslant \varphi \leqslant \pi$, the equation always has two solutions φ_1 and $\varphi_2 = \varphi_1 + \pi/2$. Therefore at every point of a general plane stress field there always exist two mutually perpendicular directions in which the shear stresses become zero. Furthermore, the derivative

$$\frac{d\sigma}{d\varphi} = (\sigma_y - \sigma_x)\sin 2\varphi - 2\tau_{xy}\cos 2\varphi = -2\tau$$

has the same zero values, i.e. the stresses in the two particular directions have extreme values and are, for this reason, referred to as principal stresses σ_1 and σ_2.

From equation (2.15) it appears that the strains also have extreme values in the principal stress directions, so that *the directions of principal strain and principal stress are always identical.*

As will be shown in the next section of this book, the directions of the principal strains may also be found experimentally without having recourse to shear stresses or shear strains.

2.2.4 DETERMINING THE PRINCIPAL STRESSES FROM STRAIN MEASUREMENTS

As we have seen, equation (2.11) describes the strain field around a point directly in the form in which it would be measured by any strain-measuring instrument—as linear changes concentrically around a point. We shall accordingly revert to that equation and write down its derivative with respect to φ; we shall now seek the zero positions:

$$\frac{2(1 + \mu)}{E}(\sigma_1 - \sigma_2)\sin \varphi \cos \varphi = 0.$$

On condition that $\sigma_1 \neq \sigma_2$ ($\sigma_1 = \sigma_2$ corresponds to the concentric circles in Figure 2.19, Case 3), the equation has single-valued solutions for $\varphi = n(\pi/2)$. From this it again appears that the extreme values for ε_φ coincide with the

directions of the principal stresses. Because of this fact, it is possible to describe the whole strain field, and hence the stress field, entirely in terms of the given principal strains. Putting (as has already tacitly been done in the preceding section) $\varepsilon_{\varphi=0} = \varepsilon_1$ and $\varepsilon_{\varphi+\pi/2} = \varepsilon_2$, we may substitute the appropriate expressions into equation (2.11) and obtain the relationship already derived:

$$\varepsilon_\varphi = \varepsilon_1 \cos^2 \varphi + \varepsilon_2 \sin^2 \varphi. \tag{2.12}$$

Often the directions of the principal stresses or principal strains are not known before a test. Thus, in practice, a common problem is the determination of the magnitude and direction of the principal strains, and therefore of the principal stresses, from a number of strain measurements in known directions.

For determining the two principal strains and the orientation angle, it is evidently necessary to have three strain measurements ε_1, ε_{II}, ε_{III} in different directions which are known and duly defined in relation to one another. The direction of ε_1 which may, for instance, be desired can then be defined with reference to one of those known directions of strain measurement.

Theoretically we are free to choose the three strain measurement directions. However, since each of the three measured values is affected by errors there is some advantage in choosing the three directions whose measurements will provide extreme values. In general, strain gauges in a rosette arrangement (Section 3.3.1) are used for determining the principal strains. With such rosettes, the angles between the three directions of strain measurement are pre-determined.

The mathematical relations for determining the principal strains will now be derived.

Figure 2.23 represents a general plane strain field. The strains ε_1, ε_{II}, ε_{III} are assumed to have been determined by measurement, while the angles ψ_2 and

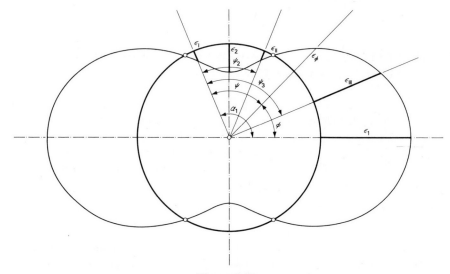

Figure 2.23

ψ_3 between them are known. Let α_1 denote the unknown angle between the strain direction ε_1 and the direction of the principal strain ε_1. We wish to determine this latter direction.

From equation (2.12), with $\varphi = \alpha_1 - \psi$, we obtain:

$$\varepsilon_\varphi = \varepsilon_1 \cos^2(\alpha_1 - \psi) + \varepsilon_2 \sin^2(\alpha_1 - \psi). \tag{2.17}$$

The three measured strains can be linked to the required co-ordinate system by means of the following conditional equations:

$$\varepsilon_I = \varepsilon_1 \cos^2 \alpha_1 + \varepsilon_2 \sin^2 \alpha_1$$

$$\varepsilon_{II} = \varepsilon_1 \cos^2(\alpha_1 - \psi_2) + \varepsilon_2 \sin^2(\alpha_1 - \psi_2)$$

$$\varepsilon_{III} = \varepsilon_1 \cos^2(\alpha_1 - \psi_3) + \varepsilon_2 \sin^2(\alpha_1 - \psi_3). \tag{2.18}$$

The three quantities ε_1, ε_2 and α_1 that we seek to determine may be calculated from these equations. If the intermediate angles ψ are given certain special values (e.g. multiples of $\pi/4$ or $\pi/6$), the solution of the set of equilibrium equations becomes simple. For this reason, strain gauge rosettes are made with angles having these values.

As an example, the equations will now be solved for a particular case, namely, a rosette with $\psi_2 = 45°$ and $\psi_3 = 90°$. Substituting these values into equation (2.18), we obtain:

$$1. \quad \varepsilon_I = \varepsilon_1 \cos^2 \alpha_1 + \varepsilon_2 \sin^2 \alpha_1$$

$$2. \quad \varepsilon_{II} = (\varepsilon_1 + \varepsilon_2) + (\varepsilon_1 - \varepsilon_2)\sin 2\alpha_1$$

$$3. \quad \varepsilon_{III} = \varepsilon_1 \sin^2 \alpha_1 + \varepsilon_2 \cos^2 \alpha_1. \tag{2.19}$$

Addition of relations 1 and 3 of equation (2.19) gives the expression for the invariance of the sum of two mutually perpendicular strains which is often of practical value:

$$\varepsilon_I + \varepsilon_{III} = \varepsilon_1 + \varepsilon_2 = 2A. \tag{2.20}$$

Subtraction of relations 1 and 3 of equation (2.19) gives:

$$\varepsilon_I - \varepsilon_{III} = (\varepsilon_1 - \varepsilon_2)\cos 2\alpha_1.$$

On substitution of these values into relation 2 of equation (2.19):

$$2\varepsilon_{II} = \varepsilon_I + \varepsilon_{III} + (\varepsilon_I - \varepsilon_{III})\tan 2\alpha_1$$

whence we obtain:

$$\tan 2\alpha_1 = \frac{2\varepsilon_{II} - \varepsilon_I - \varepsilon_{II}}{\varepsilon_I - \varepsilon_{III}}$$

and we now calculate:

$$\varepsilon_1 - \varepsilon_2 = \frac{\varepsilon_I - \varepsilon_{III}}{\cos 2\alpha_1} = (\varepsilon_I - \varepsilon_{III})\sqrt{1 + \tan^2 2\alpha_1}.$$

Substituting in equation (2.19) we obtain:

$$\varepsilon_1 - \varepsilon_2 = \sqrt{(\varepsilon_I - \varepsilon_{III})^2 + (2\varepsilon_{II} - \varepsilon_I - \varepsilon_{III})^2}$$

or:

$$\varepsilon_1 - \varepsilon_2 = \sqrt{2}\sqrt{(\varepsilon_I - \varepsilon_{II})^2 + (\varepsilon_{II} - \varepsilon_{III})^2} = 2B. \qquad (2.21)$$

The principal strains can now be determined from equations (2.19) and (2.21):

$$\varepsilon_1 = A + B$$

$$\varepsilon_2 = A - B.$$

It should be noted that the angle between the strain gauge axis for measuring the strain ε_I and the direction of the principal strain—which has been defined as positive in the anticlockwise direction in Figure 2.23—is not uniquely determined by the algebraic sign of the tangent. If there is a risk of confusion of the principal stress directions, the angle in question can be determined by applying the sign convention stated in Table 2.1 for the values of the numerator and denominator in the formula for $\tan 2\alpha_1$.

Table 2.1

Numerator	Denominator	Range
+	+	$0 < 2\alpha_1 < \dfrac{\pi}{2}$
+	−	$\dfrac{\pi}{2} < 2\alpha_1 < \pi$
−	−	$\pi < 2\alpha_1 < \dfrac{3\pi}{4}$
−	+	$\dfrac{3\pi}{4} < 2\alpha_1 < 2\pi$

For other pre-determined values of the angles between the strain gauges, the principal strains can be determined in a similar fashion. The relevant formulae applicable to common rosette arrangements are given in Figure 2.24.

If maximum accuracy is required in the use of electrical resistance strain gauges, difficulties may arise in consequence of the 'cross-sensitivity' of the gauges (Section 3.3.1).

It may occur, for example, that due to a technical defect one of the three strain measurements provided by a rosette may be very inaccurate. The whole set of measurements obtained with this rosette may then be worthless. Usually a critical appraisal of the 'plausibility' of the results will disclose any major errors, which are of rare occurrence anyway. In special cases it may be essential to be absolutely certain about the validity of the result. This can be achieved by the use of rosettes comprising more than three strain gauges, so providing extra data as a safeguard. The results of these measurements can be evaluated with the aid of the thory of least squares to check the reliability of the individual measurements.

Delta
rosette

$\psi_2 = 60°$
$\psi_3 = 120°$

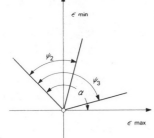

$A = \frac{1}{3}(\varepsilon_I + \varepsilon_{II} + \varepsilon_{III})$

$B = \frac{\sqrt{2}}{3}\sqrt{(\varepsilon_I + \varepsilon_{II})^2 + (\varepsilon_{II} - \varepsilon_{III})^2 + (\varepsilon_I - \varepsilon_{III})^2}$

$\tan 2\alpha = \dfrac{\sqrt{3}(\varepsilon_{II} - \varepsilon_{III})}{2\varepsilon_I - \varepsilon_{II} - \varepsilon_{III}}$

Three-element
rectangular
rosette

$\psi_2 = 45°$
$\psi_3 = 90°$

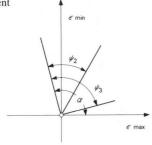

$A = \frac{1}{2}(\varepsilon_I + \varepsilon_{III})$

$B = \frac{\sqrt{2}}{2}\sqrt{(\varepsilon_I - \varepsilon_{II})^2 + (\varepsilon_{II} - \varepsilon_{III})^2}$

$\tan 2\alpha = \dfrac{2\varepsilon_{II} - \varepsilon_I - \varepsilon_{III}}{\varepsilon_I - \varepsilon_{III}}$

Four-element
rectangular
rosette

$\psi_2 = 45°$
$\psi_3 = 90°$
$\psi_4 = 135°$

$A = \frac{1}{4}(\varepsilon_I + \varepsilon_{II} + \varepsilon_{III} + \varepsilon_{IV})$

or $\frac{1}{2}(\varepsilon_I + \varepsilon_{III})$

or $\frac{1}{2}(\varepsilon_{II} + \varepsilon_{IV})$

$B = \frac{1}{2}\sqrt{(\varepsilon_I - \varepsilon_{III})^2 + (\varepsilon_{II} - \varepsilon_{IV})^2}$

$\tan 2\alpha = \dfrac{\varepsilon_{II} - \varepsilon_{IV}}{\varepsilon_I - \varepsilon_{III}}$

Tee-delta
rosette

$\psi_2 = 60°$
$\psi_3 = 120°$
$\psi_4 = 30°$

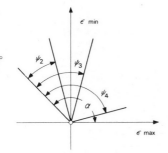

$A = \frac{1}{2}(\varepsilon_{III} + \varepsilon_{IV})$

or $\frac{1}{2}(\varepsilon_I + \varepsilon_{II} + \varepsilon_{III})$

$B = \frac{\sqrt{2}}{2}\sqrt{(\varepsilon_I - \varepsilon_{II})^2 + (\varepsilon_{II} - \varepsilon_{III})^2 + (\varepsilon_I - \varepsilon_{III})^2}$

$\tan 2\alpha = \dfrac{\sqrt{3}(\varepsilon_{II} - \varepsilon_{III})}{2\varepsilon_I - \varepsilon_{II} - \varepsilon_{III}}$

Figure 2.24   Delta rosette.

The arithmetic evaluation of the sets of formulae from data yielded by hundreds of strain gauge rosettes, each for many different loading conditions of the test model, can be laborious and time-consuming. All sorts of aids have been devised to ease this task. Nomographs are available for graphically determining the principal strains from the measured values. Simple analog computers are commercially obtainable which simulate the analytical transformations, including the stress calculation, by means of electrical circuits. The desired value of Poisson's ratio can be set by means of a rotary potentiometer and the results can be read from the indicating instruments.

These aids and devices are made superfluous by the inclusion of a digital computer in the measuring circuit (Section 4.2). The computer can carry out the necessary transformations in fractions of a second, so that it can cope with a rapid flow of incoming data. The computer can function on-line and by operating during testing does not hold up the measurement process.

2.3 DETERMINING THE INTERNAL FORCES IN ELASTIC STRUCTURES

2.3.1 STATEMENT OF THE PROBLEM

The investigation of the plane state of stress by means of strain measurements has been dealt with earlier (Section 2.2.3). In general, we can determine the stresses and strains at any point on the surface of the model, and in cases where we are only interested in the extreme fibre stresses in selected regions of the structure this solves the problem.

But if the purpose of the model test is to clarify and comprehend the load-carrying behaviour of a structure as a whole and if experimental results are to serve as a basis for its design, then it becomes necessary also to know the internal forces and moments (collectively termed 'stress resultants') which cannot be determined by direct measurement. On the assumption of some particular internal stress distribution, the problem is to deduce the stress resultants from the measured extreme fibre stresses. However, such an approach runs into practical as well as theoretical difficulties. Premature and incautious interpretations of the results of measurements can easily lead to wrong conclusions about the structural behaviour. Above all, the experimenter must on principle never trust a single measured value on its individual merits. Results must be brought into rational and intelligible relationships to each other.

The pitfall of committing *theoretical errors of interpretation* has already been referred to (Section 1.2). Calculations of stress resultants from known extreme fibre stresses, i.e. stresses acting at the outermost surface of the structure, must be based on a suitable assumed stress distribution inside the structure. Such an assumption can be reached only by considering certain hypotheses of elastic theory. Hence the accuracy of the inference about the strain resultants depends upon the validity of the simplified assumption with regard to the elastic behaviour of the model and the structure it represents. For instance, the application of Navier's hypothesis—the assumption that the stress (or strain) at any point due to bending is proportional to its distance from the neutral axis—to sections where local stress concentrations exist gives rise to interpretations which are not only quantitatively but also qualitatively wrong. An example of such incorrect interpretation of a strain measurement for determining the stress resultants at a corner of a rigid frame is given in Figure 2.25.

Since the actual stress distribution across the column of the frame is not known, it is not possible to deduce the stress resultants at this section from the strain measurement alone. The location of the points where strain measurements have to be performed in order to determine the internal forces in a structure must therefore be chosen with particular care.

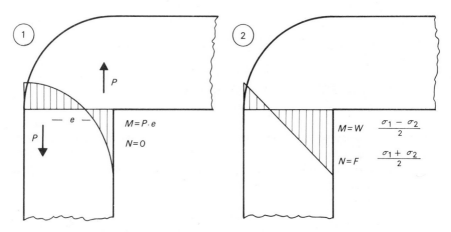

Figure 2.25   Stress distribution at the corner of a rigid frame. 1. Actual behaviour. 2. Incorrect interpretation based on Navier's hypothesis.

Practical errors of interpretation arise from unavoidable random variations in the properties of the material and the geometry of the model and from imperfections in the test set-up. When we perform measurements on a particular type of structural component (a beam, a plate, a slab, etc.), we tend to assume that the properties which we associate, by definition, with the type in question will also be present in the model. This is not necessarily so. On the contrary, factors that give rise to errors are liable to creep into the measuring arrangement, and the resulting errors may well be of the same order of magnitude as the values that we wish to determine. However, by means of carefully chosen check measurements it is usually possible to filter out the undesirable influences. A good example of what is meant is afforded by a thin plate loaded in its own plane. It is evident that unavoidable bending effects due to minor irregularities (lack of planeness) of the plate or to slight eccentricity of the load position can significantly affect the magnitude of the stresses measured in the extreme fibres of the test specimen. The effect of curvature can be separately estimated, and thus eliminated, by simultaneous measurement of the strains at points located on opposite faces of the plate.

In every structural model, there are more or less extensive regions in which the stress resultants cannot be determined with adequate accuracy merely from a limited number of measurements. It is up to the experimenter to choose the theory on which to base his assumption for the probable internal stress distribution. Meaningful planning and interpretation of the tests are therefore

largely dependent on his experinece and judgment. However, the equilibrium check (Section 3.2.2) does provide us with a useful means for obtaining objective and quantitative verification of the assumed interpretative functions at selected sections. This check, which should never be omitted, presupposes that all the reactions R_l of the model are transmitted to the base or sub-structure through dynamometers (Fig. 2.26).

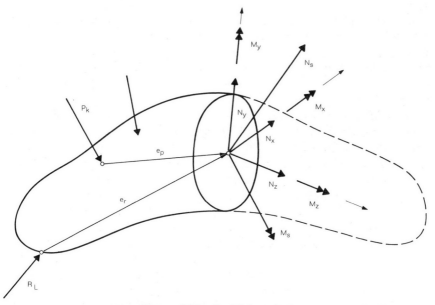

Figure 2.26 Equilibrium check.

To begin with, the reliability of the reaction measurements themselves can be verified by means of the following relation (whose existence offers a major advantage over any strain measurement):

$$\sum_1^K \vec{P}_k + \sum_1^L \vec{R}_l = 0.$$

A very large number of strain gauges are fixed around the edge of the section chosen for the equilibrium calculation and the stress resultants $\vec{N}$ and $\vec{M}$ are calculated by means of numerical integration from the stress distribution determined here, and from the assumed internal stress distribution. If the latter assumption is indeed correct, these stress resultants will have to be in equilibrium with the external forces acting to the left or right of the section, i.e.:

$$\vec{N}_s + \sum_1^s \vec{R}_l + \sum_1^s \vec{P}_l = 0$$

$$\vec{M}_s + \sum_1^s R_l e_r + \sum_1^s P_l e_r = 0.$$

The wrong interpretation illustrated in Figure 2.25, for example, would have been detected by applying such an equilibrium check. If the column of the frame had been supported on dynamometers for measuring the reaction, the absence of the normal force would have been directly apparent, the moment $M = P \cdot e$ could have been determined and it would have been possible to form a fairly accurate idea of the actual distribution of the stresses across the section.

For determining stresses and stress resultants it is generally necessary to establish an analytical inter-relation between the measured values obtained from a number of gauges. According to the particular quantity to be determined, different sets of individual measurements are thus brought into association with one another. To avoid possible confusion, the following concepts will now be defined:

(a) *Measuring position* (m): the position, or point, at which an individual measuring gauge is located. A triple rosette arrangement of strain gauges thus comprises three measuring positions, each of which therefore corresponds to a data channel in the experimental equipment.

(b) *Measuring locality* (o): the locality where a complete picture of the strains and stresses at the surface of the model is obtained by means of associated measuring positions. A rosette of strain gauges would thus constitute a measuring locality. So would a set of (say, six) bearing reaction measuring devices.

(c) *Section point* (s): the point for which, within a given section, the stress resultants are calculated from the measured data yielded by the surrounding measuring positions or localities or are brought into relationship with the external forces.

In the following treatment of the subject, the problems connected with the evaluation of model measurements will be considered with reference to examples of familiar types of structure. It will be seen that reliable results can be obtained only by the exercise of the greatest possible care in the technical construction of the test arrangement and constant awareness of the risk of incorrect interpretations. It may come as a surprise to see how easily the actual behaviour of a structure may differ considerably from that assumed in theory.

2.3.2 STRUCTURES COMPOSED OF BAR-TYPE MEMBERS (FRAMEWORKS)

With the aid of electronic computers, it is now nearly always possible to deal with such structures by direct analytical calculation. Only in very rare cases will it be meaningful to use model-testing techniques for their investigation, Such exceptions are more particularly those framed structures which constitute sub-assemblies of a complex overall structure and whose boundary conditions are therefore not known in advance. Examples include the beam-type elements into which slabs and similar structures are sometimes assumed

to be sub-divided for the purpose of analysis; also, connecting elements between stressed-skin structures, flexurally rigid elements in suspension structures, etc.

Framed structures are composed of beams of known cross-sectional dimensions and properties for which Navier's hypothesis is assumed to be
valid. In general, the loading conditions of such members consist of shear
force, normal force, biaxial bending and torsion.

With reference to the rectangular beam, it will first be shown how the stress
resultants, which are determined by six quantities, can be found by means of
strain measurements. In Figure 2.27, the stress distribution patterns (as a
consequence of unit stresses) associated with the various stress resultants are
shown. These patterns, each provided with an appropriate proportionality
factor, must be imagined superimposed upon one another at a given section
point. The strains associated with Q_x and M_x and with Q_y and M_y at F_y and F_x,
respectively, are zero and therefore not indicated diagrammatically.

It is seen that the principal strains (in the axial direction of the beam) due
to M and N are unaffected by those due to Q and M_T, whereas the principal
strains (at 45° to the axis) due to Q and M_T comprise a proportion of strain
due to normal and flexural stress. The last-mentioned strains can therefore
always be determined first and separately.

Theoretically, the strain distribution can be determined by measurements
performed at three arbitrarily chosen points on the surface of the beam. For
practical reasons, however, it will be advantageous to choose three specially
situated points, e.g. at the corners of the beam cross-section or, as in the following derivation, at the mid-points of the sides $(-b/2, +b/2, h/2)$. Then:

$$\varepsilon^{-\frac{1}{2}b} = \frac{1}{E}\left(\frac{N}{F} + \frac{M_y}{W_y}\right)$$

$$\varepsilon^{+\frac{1}{2}b} = \frac{1}{E}\left(\frac{N}{F} - \frac{M_y}{W_y}\right)$$

$$\varepsilon^{+\frac{1}{2}h} = \frac{1}{E}\left(\frac{N}{F} + \frac{M_x}{W_x}\right) \tag{2.22}$$

whence we obtain:

$$N = \tfrac{1}{2}EF(\varepsilon_0^{-\frac{1}{2}b} + \varepsilon_0^{+\frac{1}{2}b})$$
$$M_x = \tfrac{1}{2}EW_x(\varepsilon_0^{-\frac{1}{2}b} + \varepsilon_0^{+\frac{1}{2}b} - 2\varepsilon^{\frac{1}{2}h})$$
$$M_y = \tfrac{1}{2}EW_y(\varepsilon_0^{-\frac{1}{2}b} - \varepsilon_0^{+\frac{1}{2}b}). \tag{2.23}$$

The subscript 0 refers to the direction of the strain gauge (angle 0 with the
axis of the beam). The strains $\varepsilon_{\pi/4}$ must first have the strain ε_0 due to normal
stress deducted from them in order to be regarded as the shear strain ε_1 (see

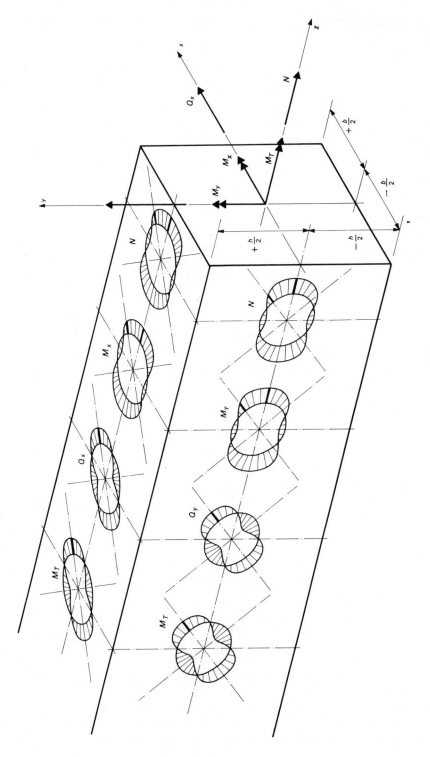

Figure 2.27

Case 1 in Figure 2.19):

$$\varepsilon_1^{-\frac{1}{2}b} = \varepsilon_{\pi/4}^{-\frac{1}{2}b} - \tfrac{1}{2}(1 - \mu)\varepsilon_0^{-\frac{1}{2}b}$$

$$\varepsilon_1^{+\frac{1}{2}b} = \varepsilon_{\pi/4}^{+\frac{1}{2}b} - \tfrac{1}{2}(1 + \mu)\varepsilon_0^{+\frac{1}{2}b}$$

$$\varepsilon_1^{+\frac{1}{2}h} = \varepsilon_{\pi/4}^{+\frac{1}{2}h} - \tfrac{1}{2}(1 + \mu)\varepsilon_0^{+\frac{1}{2}h}. \tag{2.24}$$

The tacit assumption that despite the finite size of the strain gauge the strain $\varepsilon_{\pi/4}^{-\frac{1}{2}b}$ remains unaffected by the strain distribution due to the moment M_x is indeed correct if the measurement is performed at the neutral axis, since this latter strain becomes zero when integrated over the gauge length.

Bearing in mind that

$$Q = \frac{2}{3}FG\gamma = \frac{2}{3}\frac{EF}{(1 + \mu)}\varepsilon_1^Q$$

$$M_T = W_T G\gamma = \frac{EW_T}{(1 + \mu)}\varepsilon_1^T$$

and since $W_T^{\frac{1}{2}b} = W_T^{-\frac{1}{2}b}$, we obtain for our beam:

$$\varepsilon_1^{-\frac{1}{2}b} = \frac{(1 + \mu)}{E}\left(\frac{3}{2}\frac{Q_y}{F} + \frac{M_T}{W_T^{\frac{1}{2}b}}\right)$$

$$\varepsilon_1^{+\frac{1}{2}b} = \frac{(1 + \mu)}{E}\left(\frac{3}{2}\frac{Q_y}{F} - \frac{M_T}{W_T^{\frac{1}{2}b}}\right)$$

$$\varepsilon_1^{+\frac{1}{2}h} = \frac{(1 + \mu)}{E}\left(\frac{3}{2}\frac{Q_x}{F} + \frac{M_T}{W_T^{+\frac{1}{2}h}}\right) \tag{2.25}$$

whence we obtain the stress resultants:

$$Q_x = \frac{EF}{3(1 + \mu)}\left[2 \cdot \varepsilon_1^{+\frac{1}{2}h} - \frac{W_T^{\frac{1}{2}h}}{W_T^{\frac{1}{2}h}}(\varepsilon_1^{-\frac{1}{2}b} - \varepsilon_1^{+\frac{1}{2}b})\right]$$

$$Q_y = \frac{1}{3}\frac{EF}{(1 + \mu)}(\varepsilon_1^{-\frac{1}{2}b} + \varepsilon_1^{+\frac{1}{2}b})$$

$$M_T = \frac{EW_T^{\frac{1}{2}b}}{2(1 + \mu)}(\varepsilon_1^{-\frac{1}{2}b} - \varepsilon_1^{+\frac{1}{2}b}). \tag{2.26}$$

It has so far been assumed that the section properties of the beam under consideration, and more particularly the shear stresses due to torsion at the measuring positions, are known. This latter assumption is not necessarily valid for a beam of arbitrary cross-sectional shape, however. Let us consider such a beam. For determining the stress resultants we can proceed as follows:

First, the principal axes of the cross-section are determined analytically. Then, as in the above example, the longitudinal strains ε_0 are measured at the three points of intersection of these axes with the edges of the cross-section. With the strain values thus determined, the stress resultants N, M_x, M_y can be calculated.

At the points where the longitudinal strain measurements are obtained, we again determine the strains $\varepsilon_{\pi/4}$ and reduce them to the values ε_1 in accordance with equation (2.11). Since the torsional section modulus W_T is not known, it is not possible directly to separate the shear force contribution to the 'diagonal' strain from the torsional contribution.

This difficulty can be overcome by means of additional measurements. In a portion of the (straight) bar without external load, the normal force and the torsion remain unchanged. Only the bending moments vary in the z direction and are linked to the shear force by the relations:

$$Q_x = -\frac{dM_y}{dz} \quad \text{and} \quad Q_y = -\frac{dM_x}{dz}.$$

The shear force can therefore be determined from the change in the bending moment as found by performing strain measurements at two points spaced a distance Δz apart in the z direction. Since the normal force remains constant, two additional measurements are sufficient. The shear force is then:

$$Q_x = -\frac{M_y(z + \Delta z) - M_y(z)}{\Delta z}; \qquad Q_y = -\frac{M_x(z + \Delta z) - M_y(z)}{\Delta z}.$$

The principal 'diagonal' strains due to this shear force can now be calculated for the corresponding measuring points and deducted from the strains ε_1 measured at these points. The remaining strain is attributable purely to torsion. The magnitude of the torsional moment itself can be determined by means of Prandtl's membrane analogy for which purpose the local shear stresses determined are used for normalising the boundary conditions.

2.3.3 SLABS

2.3.3.1 Theory and actual behaviour

We shall measure strains (or curvatures) on a slab or plate, a flat structure whose dimensions in two directions (length and width) are many times larger than its dimension in the third direction (thickness h) and which is loaded perpendicularly to its own plane. We wish to deduce the stress resultants from the strain measurements. In order to do this we shall employ the relation between (external) strain and (internal) stress resultants expressed by Kirchhoff's theory of plates. If this method of dealing with the problem is to be successful, it will be necessary to consider two questions:

(i) Does the test set-up (support and loading arrangements) provide a sufficiently accurate simulation of the limiting conditions of the theory adopted as the basis of comparison and interpretation?

(ii) At which points of the model *must* the strains differ from those expected according to the theory and must therefore not be used for calculating the stress resultants?

Besides the general hypotheses of elastic theory, the hypotheses more particularly applicable to the theory of slabs are:

(a) Navier's hypothesis for the stress (and strain) distribution in the depth (thickness) of the slab.

(b) Small dimensions in the z direction (thickness).

(c) Small deflections in comparison with the slab thickness, hence no strains in the middle surface of the slab.

(d) That the effect of shear stresses in the z direction upon the deformation is negligible.

In the following treatment of the problem the limits for the application of these assumptions will be quantitatively investigated. The relationships between (extreme fibre) strains or curvatures and stress resultants are:

$$m_x = k\left(\frac{\partial^2 w}{\partial x^2} + \mu\frac{\partial^2 w}{\partial y^2}\right)$$

$$m_y = k\left(\frac{\partial^2 w}{\partial y^2} + \mu\frac{\partial^2 w}{\partial x^2}\right)$$

$$m_{xy} = (1 - \mu)k\frac{\partial^2 w}{\partial x \partial y} \qquad (2.27)$$

where $k = Eh^3/[12(1 - \mu^2)]$ denotes the stiffness of the slab. These relationships will be used for evaluating the strain measurements performed on the slab.

Assuming small deflections and therefore small slopes, the expressions $\partial^2 w/\partial x^2$ and $\partial^2 w/\partial y^2$ can be equated to the curvatures $1/r_x$ and $1/r_y$ respectively. The expression $\partial^2 w/\partial x \partial y$ describes the 'twist' of the element. It is therefore possible to deduce the moments acting in the slab from curvature and torsion measurements.

Leonhard and Andrä have developed a commercially available curvature-measuring device which takes account of the influence of Poisson's ratio by electrical coupling the inductive curvature measurements in two mutually perpendicular directions. It gives direct indications of the moments to be determined. Although this method may be useful for special purposes (direct graphical recording of influence surfaces), we are more particularly concerned with automation and numerical processing of the measurements obtained, and for this reason the technique of curvature measurement will not be considered in further detail.

We shall start from the strains, which are related to the curvature and twist as follows:

$$\varepsilon_x = \frac{h}{2}\frac{\partial^2 w}{\partial x^2}; \qquad \varepsilon_y = \frac{h}{2}\frac{\partial^2 w}{\partial y^2}; \qquad \gamma_{xy} = h\frac{\partial^2 w}{\partial x \partial y}$$

With these expressions, the equations expressed in (2.27) above become:

$$m_x = \frac{2k}{h}(\varepsilon_x + \mu\varepsilon_y) = \frac{Eh^2}{6(1 - \mu^2)}(\varepsilon_x + \mu\varepsilon_y)$$

$$m_y = \frac{2k}{h}(\varepsilon_y + \mu\varepsilon_x) = \frac{Eh^2}{6(1 - \mu^2)}(\varepsilon_y + \mu\varepsilon_x)$$

$$m_{xy} = (1 - \mu)\frac{k}{h}\gamma_{xy} = \frac{Eh^2}{12(1 + \mu)}\gamma_{xy} = \frac{Eh^2}{6(1 + \mu)}\varepsilon_{\pi/4}. \qquad (2.28)$$

We have thus established the relations linking plane strain measured at the slab surface and the stress resultants. Introducing the section modulus of the slab per unit length and bearing in mind that:

$$m_x = \sigma_x\frac{h^2}{6}; \qquad m_y = \sigma_y\frac{h^2}{6}; \qquad m_{xy} = \tau_{xy}\frac{h^2}{6}$$

the equations in (2.28) can be transformed to the well-known expressions relating to plane stress:

$$\sigma_x = \frac{E}{1 - \mu^2}(\varepsilon_x + \mu\varepsilon_y)$$

$$\sigma_y = \frac{E}{1 - \mu^2}(\varepsilon_y + \mu\varepsilon_x)$$

$$\tau_{xy} = \frac{E}{2(1 + \mu)}\gamma_{xy} = G\gamma.$$

If we know that our model slab does indeed function as a true plate, it is unnecessary to investigate the conditions existing at points within its thickness. The stress distribution at one of the surfaces of the slab will then differ from the internal stress distribution merely by the proportionality factor $h^2/6$.

2.3.3.2 Measurement of boundary conditions

As mentioned above, the stress resultants in a slab can always be determined from strain measurements in three known directions at the surface of the slab. At the edges of the slab, where the stress resultants are always of particular interest, it is generally possible to determine these from fewer than three measuring positions per measuring locality if special bearing conditions are satisfied whereby certain features of the stress or strain patterns can be directly inferred. The considerations presented below are summarised in Figure 2.28.

At the *restrained edge* (parallel to the y direction), both the curvature $\partial^2 w/\partial y^2$ and the twist $\partial^2 w/\partial x\,\partial y$, and therefore ε_y and γ_{xy}, are zero. The restraint

Rigidly restrained (clamped) edge

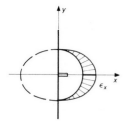

$$m_x = \frac{1}{1 - \mu^2} \frac{Eh^2}{6} \varepsilon_x$$

$$m_y = \frac{\mu}{1 - \mu^2} \frac{Eh^2}{6} \varepsilon_x$$

$$m_{xy} = 0$$

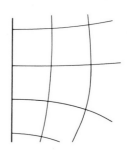

Freely rotatable edge on immovable base

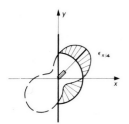

$$m_x = 0$$

$$m_y = 0$$

$$m_{xy} = \frac{1}{1 + \mu} \frac{Eh^2}{6} \varepsilon_{\pi/4}$$

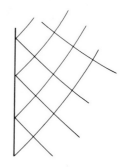

Unsupported or elastically supported edge

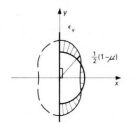

$$m_x = 0$$

$$m_y = \frac{Eh^2}{6} \varepsilon_y$$

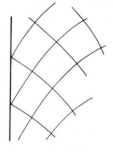

$$m_{xy} = \frac{1}{1 + \mu} \frac{Eh^2}{6} (\varepsilon_{(\pi/4)} - \tfrac{1}{2}(1 - \mu)\varepsilon_x)$$

Figure 2.28   Boundary conditions for a slab.

moments can therefore be determined by measuring ε_x only, as follows:

$$m_x = \frac{1}{1 - \mu^2}\frac{Eh^2}{6}\varepsilon_x$$

$$m_y = \frac{\mu}{1 - \mu^2}\frac{Eh^2}{6}\varepsilon_x$$

$$m_{xy} = 0.$$

This boundary condition corresponds to Case 5 in Figure 2.21.

At the *freely supported edge* (rotatably resting on a *rigid* support), the restraint moment $m_x = 0$. Therefore $\varepsilon_x = \mu\varepsilon_y = 0$. Since the rigid support along x remain straight, the curvature $\partial^2 w/\partial y^2$ and therefore ε_x must also be zero. Therefore:

$$\varepsilon_x = \varepsilon_y = 0$$

and

$$m_x = m_y = 0.$$

Hence only a twist can occur, which is measured at an angle $\pi/4$ to the edge:

$$m_{xy} = \frac{1}{(1 + \mu)}\frac{Eh^2}{6}\varepsilon_{\pi/4}.$$

Hence for these support conditions also, one measurement is sufficient to determine the three internal moments.

At the *unsupported edge*, or the edge freely supported on a *yielding* support, there is no restraint moment, so that here too we have $\varepsilon_x + \mu\varepsilon_y = 0$. Because of this relation, we can content ourselves with measuring either ε_x or ε_y. We shall measure ε_y since its value is generally more significant, and thus obtain:

$$m_y = \frac{Eh^2}{6}\varepsilon_y.$$

This corresponds to a uni-axial state of stress (Case 5 in Figure 2.19). From the assumption, we have:

$$m_y = 0.$$

The twist is measured at an angle $\pi/4$ to the edge. It leaves the measured value for the strain ε_y unaffected. On the other hand, the strain measured in the 45° direction includes, in addition to the twist effect, a proportion of strain which is due to ε_y and must therefore be eliminated.

The measured total strain in the direction $\pi/4$ is (see Cases 2 and 4 in Figure 2.19):

$$\varepsilon_{\pi/4} = \frac{1}{2}(1 - \mu)\frac{\sigma_y}{E} + (1 + \mu)\frac{\tau_{xy}}{E}$$

or

$$\varepsilon_{\pi/4} = \frac{1}{2}(1 - \mu)\varepsilon_y + (1 + \mu)\frac{6}{Eh^2}m_{xy}$$

whence we obtain for the twisting moment:

$$m_{xy} = \frac{1}{(1 + \mu)}\frac{Eh^2}{6}\left[\varepsilon_{\pi/4} - \frac{1}{2}(1 - \mu)\varepsilon_y\right].$$

The boundary values for this loading case can be determined from measurements obtained at two measuring positions.

From the theoretical point of view, there is apparently nothing to stop us measuring the stress resultants at the slab edge. However, anyone who has had practical experience of technical measurements on slabs knows how difficult it is to simulate the boundary conditions which are so simple to formulate analytically.

The problem of producing a continuously supported and yet free rotatable edge is beyond the possibilities of a normal properly equipped mechanical workshop. Instead, knife-edge rocker bearings have to be used as a substitute, although this significantly alters the state of stress along the edge of the slab. In addition, in actual practice the slab has to cantilever at least some minimum distance beyond the theoretical line of support, and this also affects the boundary conditions.

Full edge restraint, i.e. complete rigidity by clamping with a definitely determined line of restraint, can only be realised approximately.

Only the unsupported edge presents no difficulties from the viewpoint of technical execution. But here we encounter difficulties of measuring technique which are bound up with plate theory (Section 2.3.3.3).

2.3.3.3 Shear stress and twist

As in beam theory, we neglect the shear deformation due to the action of shear forces in plate theory also. This involves a theoretical contradiction since the shear stresses are conversely determined from the pure flexural deformation. Additionally in the plate the shear forces are in equilibrium with the twisting moments.

The shear-stress distribution due to twisting, or torsion, like that due to bending, is assumed to increase linearly from the middle to the exterior surface of the slab. For reasons of infinitesimal equilibrium, this linear distribution is not possible at the edge of the slab however, because here the external shear stresses in the desired distribution are absent. The postulated shear stress distribution must therefore have an opportunity to build up within the material in the edge zone of the slab.

The order of magnitude of the deviation of the actual elastic behaviour from that assumed in plate theory can be estimated with the aid of a simple example (Fig. 2.29).

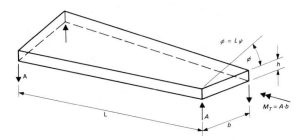

Figure 2.29   Torsion of a flat bar or slab.

Consider a slab, of thickness h and width b, subjected to pure torsional loading by a twisting couple $P \cdot b$. This resembles the problem of torsion in a very flat bar. For this case the exact elastic solution is provided by Saint Venant's theory. The shear flow distribution in both cases is indicated in Figure 2.30.

The relation between twisting moment and twist is, according to plate theory:

$$m_{xy} = (1 - \mu)k\frac{\partial^2 w}{\partial x\, \partial y} = (1 - \mu)k\psi$$

The plate stiffness $k = Eh^2/[12(1 - \mu^2)]$ can, on introducing $G = E/[2(1 + \mu)]$, be written alternatively as $k = Gh^3/[6(1 - \mu)]$; and since $2m_{xy} = A$ (see Figure 2.31), we have $M_T = 2m_{xy} \cdot b$ or:

$$M_T = \tfrac{1}{3}Gh^3 b\psi^{\text{plate}}.$$

We have thus obtained a purely torsional relation from the plate equation. The Saint Venant torsional equation for the thick bar has the same form, except that it contains a correction factor k_1:

$$M_T = k_1 Gh^3 b\psi^{\text{bar}} = GJ_1^T\psi^{\text{bar}}.$$

k_1 can be calculated from

$$k_1 = \frac{1}{3}\left(1 - 0.630\frac{h}{b} - 0.052\frac{h^5}{b^5}\cdots\right)$$

For $b/h \to \infty$, the exact solution and the solution for the plate will converge, as is indeed only to be expected. In Figure 2.32 the twist determined from the

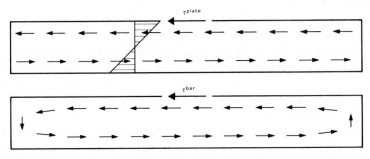

Figure 2.30   Shear flow.

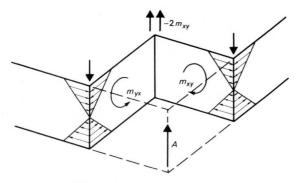

Figure 2.31 Corner of the slab.

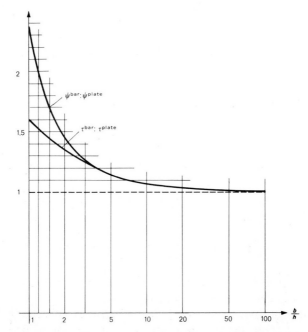

Figure 2.32 Comparison of twisting and edge shear stress according to Saint Venant and slab theory respectively.

more exact theory and the twist determined for the slab are compared. It is seen that even for a relatively very thin slab, e.g. one whose thickness is $\frac{1}{20}$ of its width, plate theory involves quite a substantial error of about 5%. Besides the twist, the relation between the shear stress on the flat face of the bar and the edge shear stress according to plate theory is also illustrated in Figure 2.32.

In very slender slabs, the effect of the simplified assumption for the edge shear flow is of no consequence to the overall behaviour of the slab. It does, however, have consequences with regard to measuring technique.

According to plate theory, the shear stresses must attain extreme values along the arrises of the plate edges and thus provide a reason for the local twisting action. In actual fact, however, careful investigation shows these shear stresses

to be zero. Hence a measurement obtained at the arris gives no information on the torsion there. We are thus compelled to consider the question: how wide is the disturbed zone outside which the vertical shear flow becomes transformed into a meaningful horizontal shear flow?

For this purpose we shall investigate the shear-stress distribution at the surface of a bar subjected to a twisting action (Fig. 2.33). We can apply an expansion into series derived from membrane analogy to this case:

$$\tau_{xy} = hG\psi - \frac{8hG\psi}{\pi^2} \sum_{n=1,3,5} \frac{1}{n^2} \frac{\cosh\left(n\pi\frac{y}{h}\right)}{\cosh\left(\frac{n\pi b}{2h}\right)}.$$

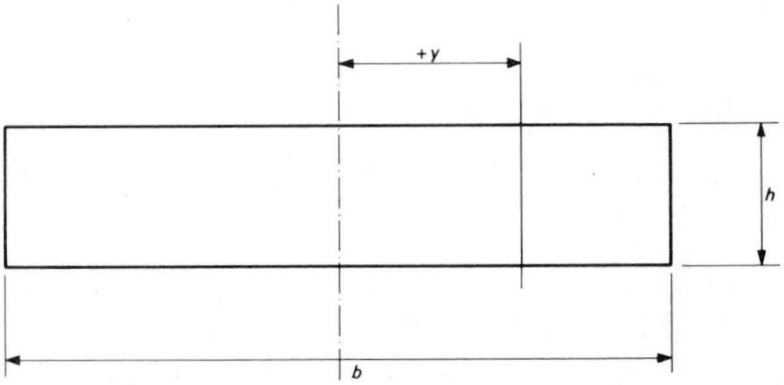

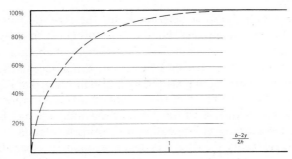

Figure 2.33   Shear-stress distribution at the surface of a flat bar according to Saint Venant.

Since we can assume $y/h > 5$ we can put $\cosh x \cong e^x$, and the above expression thus takes on the form:

$$\tau_{xy} = hG\psi \left(1 - \frac{8}{\pi^2} \sum_{n=1,3,5} \frac{1}{n^2} e^{-(b-2y)/2h} \right)$$

where the exponent $(b - 2y)/2h$ denotes the relative distance from the edge of the slab (relative to the slab thickness).

A numerical illustration of the above relation is given in Figure 2.34. From this it emerges that the shear stress builds up rapidly and reaches at

Figure 2.34   Distribution of shear stresses at the surface in the vicinity of the edge of the slab.

least 95 % of its nominal value at a distance equal to a slab thickness from the edge. Hence as a rule gauges should be installed at a distance of about 1.5 times the thickness from the slab edge in order to obtain torsional measurements.

2.3.3.4 Point supports

According to plate theory, the bending moment surface under a concentrated load (or over a point support) must correspond to an infinitely large moment at the point of application. In reality there is no such thing as an ideal point bearing. Every load, however concentrated, is applied through a contact area of finite magnitude. In addition, the plate itself is a slab of finite thickness, so that there is always some load distribution within it. The true magnitude of the support moment over a narrow bearing will depend very largely on the actual size of the bearing surface and on the thickness of the slab.

The measurement of local peak values of bending moments by means of strain or curvature measurements runs into insurmountable difficulties. For one thing, because of the length of the gauges themselves it is never possible to precisely determine sharply localised maximum values of strain. The value obtained by measurement will therefore depend on the measuring technique employed. The interpretation of the strain measurements is rendered even more dubious by the fact that the strains can be measured only on the face of the slab opposite to that at which the load is applied. Because of the decidedly three-dimensional state of the stress in the region where the force is applied the assumption that the neutral axis is located half-way through the slab is not valid in this case. In other words, the bending moments occurring over 'point' supports cannot be determined with reasonable accuracy by experimental methods.

Since the stresses and moments over the supports are the absolute maximum values that occur anywhere in a slab model, the disappointing conclusion arrived at in the last paragraph appears to make model testing meaningless in many cases. However, the importance of knowing the magnitude of the peak stress values is often over-rated. Whereas in a beam the support moment is essential in establishing the equilibrium of the system, in a slab the moment over a point-type support is merely of local significance. Adjacent to such a support the moment diminishes very rapidly in all directions, and for the equilibrium of the system only the integral of the 'support moments' for all the bearings is what matters. Thus overall equilibrium can be achieved even if the actual values of the stresses over a support vary over a wide range. Also, in the actual structure a certain amount of local plastic deformation or (more particularly in reinforced concrete) minor cracking will occur, so that the 'true' value of the support moment can never be accurately ascertained.

In reality the supports, e.g. reinforced concrete columns, always comprise a bearing area of finite extent. In spite of the above-mentioned difficulties, the following method of measuring support moments has been found successful in model testing.

The column is simulated by an annular bearing. This sort of bearing does indeed provide a very realistic approximation to the actual conditions of transmission of force from a circular column into a slab, for because of the curvature of the slab over the head of the column, the normal stresses are in fact concentrated around the periphery of the column. The diameter of the ring used as the annular support in the model may be 80% of the column diameter.

The annular bearing also offers major advantages in terms of measuring technique. It is possible to measure the stresses both on the top and on the bottom face of the slab. The moment inside the ring is practically constant and independent of direction, so that the measured result is unaffected by the positioning of the gauges. This can be readily demonstrated for a circular slab (of radius a) which is loaded and supported as indicated in Figure 2.35.

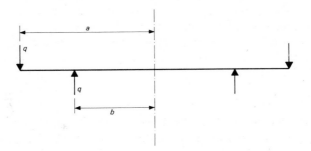

Figure 2.35   Circular slab with circular loading.

The deflected surface of such a slab carrying a distributed load of total magnitude $R = 2\pi bq$ at its edge is, within the annular bearing (of radius b), expressed by the following relation as a function of the variable radius r:

$$w = \frac{R}{16\pi k}\left[\frac{3 + \mu}{1 + \mu}(a^2 - b^2) - \frac{1 + \mu}{1 - \mu}(a^2 - b^2)\frac{r^2}{a^2} + 2b^2\left(1 + \frac{r^2}{b^2}\right)\ln\frac{b}{a}\right].$$

On calculating the moment distribution from:

$$m_r = -k\left(\frac{\partial^2 w}{\partial r^2} + \frac{\mu}{r}\frac{\partial w}{\partial r}\right)$$

$$m_\varphi = -k\left(\mu\frac{\partial^2 w}{\partial r^2} + \frac{1}{r}\frac{\partial w}{\partial r}\right)$$

we obtain:

$$m_r = m_\varphi = \frac{Q}{8\pi}\left[(1 - \mu)\frac{a^2 - b^2}{a^2} + (1 + \mu)\ln\frac{b}{a}\right] = \text{constant}.$$

The loading on the annular bearing has no significant effect and is therefore neglected. In reality, the loading around the periphery of the column is not centrally symmetric. Experiments performed with the type of annular bearing

illustrated in Figure 2.36 have shown, however, that as a result of the distributing effect of the vertically immovable edge the above assumptions as to the conditions inside the ring remain valid even if the loading is asymmetric. The flanged upper edge of the very thin-walled (light-gauge) bearing is bonded (with an adhesive) to the model slab. The cylindrical part and the flange are slotted in order to obviate annular stresses that could have a stiffening effect.

Figure 2.36 Annular bearing for simulating the effect of a point-type support.

2.3.3.5   Membrane effects

Plate theory assumes that on deflection of the slab all the points on its middle surface move in the z direction. The changes in length between two adjacent points due to the slope of the deflected slab, and the associated extensional deformation of the middle surface, are neglected. In a slab model, however, membrane stresses will inevitably occur. A distinction must be made between two effects of these stresses, both manifesting themselves as disturbances in the results of the measurements:

(i) The membrane stresses are superimposed on the extreme fibre stresses due to bending, so that the presence and magnitude of the membrane stresses are detectable when strain measurements are made at corresponding points on the two opposite faces of the slab.

(ii) The membrane stresses cause a curvature which changes the structural behaviour of the slab (flexurally rigid membrane) and hence they affect not only the normal stresses but also the distribution of bending moments which we wish to determine.

The presence of membrane stresses cannot be detected from curvature measurements, but as long as the second of the above-mentioned effects does not obtrude, such measurements (see Figure 2.39) are nevertheless more accurate than strain measurements performed merely on one side of the slab.

The membrane stresses show a certain similarity to the normal stresses occurring in the one-dimensional problem of the beam. Yet the two cases

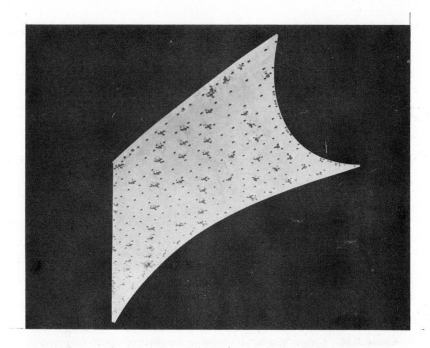

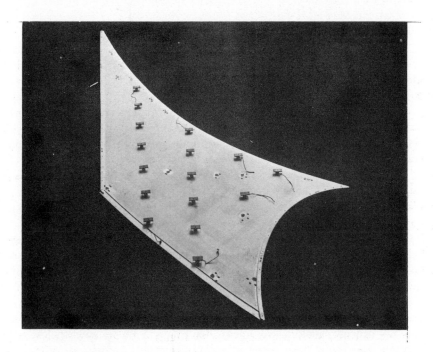

Figures 2.37 and 2.38   Top and underside views of a freely shaped bridge deck slab on point-type supports (Author's laboratory).

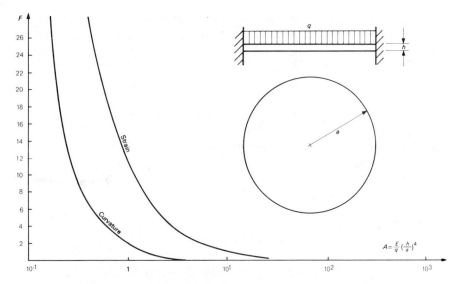

Figure 2.39 *Strain* and *curvature* errors due to membrane stresses.

differ in one important respect associated with testing techniques. Whereas in the beam to avoid the occurrence of normal stresses it is sufficient to provide bearings capable of frictionless horizontal movement, this arrangement will not by itself produce the desired result in the case of a slab. Within the slab the membrane stresses can be self-equilibrating, i.e. maintaining equilibrium with one another, so that they will occur despite free horizontal mobility of the bearings.

If we are to carry out tests on purely slab-type structures despite the impossibility of completely obviating the membrane stresses, we must consider the criteria which must be fulfilled as regards the choice of material, geometry and loading of the slab in order to keep the influence of the membrane action within prescribed permissible limits (e.g. 1% of the maximum extreme fibre strain). Because of the many conceivable shapes of slab and types of bearing, it is not possible to lay down any precise criteria for this. All that can be done is to establish a few practical rules.

For estimating the permissible test conditions, a model of freely chosen shape will be considered in relation to the circular slab analysed as a 'membrane plate' in accordance with second-order theory. For a slab supported all round its perimeter or a slab on a large number of individual supports, it is always possible to estimate an ideal radius for a circular slab which, in terms of the stress conditions acting in it, will behave approximately as the actual slab under investigation. The ratios of the curvature and the extreme fibre stress between the restrained slab analysed as a flexurally rigid membrane and the actual slab can be calculated as functions of the chosen loading q, the modulus of elasticity E and the slenderness ratio h/a. The discrepancy between the results of the two methods of analysis is plotted for curvature and for strain

in Figure 2.39. It appears that the dimensionless quantity $A = E/q\,(h/a)^4$ provides a criterion for the occurrence of membrane effects. As was to be expected, the slenderness ratio is found to be of dominant importance. For example:

A 3 mm thick aluminium plate $(E = 7 \times 10^5 \text{ kg cm}^{-2})$ of radius $r = 30$ cm is loaded with $7 \text{ kg cm}^{-2} = 7 \times 10^{-2} \text{ kg cm}^{-2}$. With these values we obtain $A = 10^{-1}$.

For this plate, the membrane effect on the measured value is about as great as the required bending effect and is therefore quite unacceptable. If the plate thickness is doubled, i.e. made equal to 6 mm, we obtain $A = 1.6 \times 10^1$. As far as the curvature measurement is concerned, the membrane effect is now of no consequence, and amounts to only 1% in the case of extreme fibre strain (or stress).

2.3.3.6   Slabs of variable thickness

The analysis of slabs or plates of variable thickness presents no difficulties in principle. However, the theory tacitly assumes that the middle surface of such a slab is in fact a plane surface, i.e. at all points exactly midway between the top and the bottom surface of the slab. In slabs of variable thickness, however, this geometric condition is seldom satisfied. Instead, as a rule, the top surface of the slab will be flat in order to carry traffic loads (as on a bridge) or sloping to ensure rainwater run-off (as on a roof).

Although it is accepted practice to analyse slabs with such a 'one-sided' variation of thickness with the aid of the usual differential equation, it should be noted that this does involve errors which are difficult to estimate and may often be quite considerable.

Since a structural model must conform to the true physical behaviour, it is always necessary to perform measurements on both sides of a slab of variable thickness in order to determine the stress resultants. In this way the neutral axis can be determined for individual strains.

Slabs with a one-sided variation of thickness are not strictly slabs within the scope of the definition. They are, indeed, to be regarded as flat shells in which the state of stress is composed of normal stresses (associated with membrane action) and flexural stresses. Through systematic strain measurements on both faces of the slab, it is possible to separate the strains ε_N due to normal stress from the flexural strains ε_B:

$$\varepsilon_B = \frac{\varepsilon^0 + \varepsilon^u}{2}$$

$$\varepsilon_N = \frac{\varepsilon^0 - \varepsilon^u}{2}.$$

The distance e from the point of zero strain to the surface of the slab of thickness h is:

$$e = \frac{h\varepsilon^0}{\varepsilon^0 - \varepsilon^u}.$$

This distance not only varies from one point to another, but also varies with the direction at the section considered since ε_N and ε_B correspond to two states of stress entirely independent of each other. The concept of the neutral axis, which is often applied to these slab-type structures in analogy with conventional beam theory, is really quite meaningless here.

The familiar mushroom floor affords an example of how little the actual structural behaviour of long-established constructional forms is understood by most designers. Experimental research on such floors with wide shallow column heads has shown that the mid-span moments as well as the moments over the supports are greatly over-estimated by theory. On the other hand, considerable tensile membrane stresses occur at the edges of the whole slab. The slab acts as an eccentric ring around the column head and thus prevents its downward deflection. In the internal bays of the floor, the hoop stresses practically cancel one another, while the edge of the whole mushroom floor has to take up the tensile membrane stresses (Fig. 2.40).

Figure 2.40 Investigation of how the size of the column head, or drop, effects the behaviour of a mushroom slab. With complete symmetry in geometry and layout of the measuring positions, it was possible to draw definite conclusions about the shell-type structural behaviour of the slab. Freely-rotatable edge support conditions were simulated in the model by a row of closely spaced point-type bearings (Author's laboratory).

2.3.4 GENERAL TYPES OF STRUCTURE

In considering the difficulties associated with measurements on slabs or plates, we have had occasion to touch on most of the problems encountered

in connection with the determination of the stress resultants in plate- and shell-type structures generally. From the viewpoint of the measurement and evaluation of the results thus obtained, there is no difference between the 'shell' and the 'plate' or 'slab' in which membrane stresses are acting.

The greatest difficulties in connection with the unambiguous determination of stress resultants always occur when the strain measurements for establishing the principal stress resultants are significantly affected by secondary actions. A typical example of this is the problem of determining the moments in the beams of a T-beam grillage, i.e. a system of intersecting beams each connected to the slab which functions as their combined top flange. As a substantial proportion of the longitudinal strain measured over a beam is due to secondary bending of the slab, these strain measurements cannot validly be used for calculating the longitudinal stresses. On the other hand, the longitudinal strain measured in the middle of the underside of the beam is solely due to the principal bending moment: no torsional strains and no strains arising from any bending moment acting in the horizontal direction are measured here. Two more measurements, performed at two opposite points on the sides of the beam, enable the neutral axis to be determined and thus enable the calculation of the triangular tensile stress distribution (within the depth of the beam) whose resultant force must, if there is pure bending, be in equilibrium with the resultant force developed in the compressive zone of the T beam. Because of the transmission of forces through the top slab within its own plane, it is likely that a normal force is transmitted into the beam; the magnitude of this normal force will vary from one loading case to another. Hence the position of the neutral axis will also vary according to the loading case concerned. From this example, the laborious nature of the optimal and realistic evaluation of experimental results is again apparent. An exhaustive analysis of the structural action is only possible through the use of a computer.

It is outside the scope of this treatment of the subject to try to deal with the vast number of possible problems associated with the evaluation of measurements. From the examples given here, it should be obvious that the investigator must proceed with great caution and with the instincts of a detective if his model experiments are to yield the information required from them.

2.3.5 POISSON'S RATIO

Poisson's ratio is a dimensionless physical constant. To satisfy the similarity conditions it must therefore have the same value in the model as in the prototype. In elastic-model tests, in which the material used for the model is usually different from the material of the actual structure, this requirement is seldom fulfilled. It is in fact the *only* requirement of similarity mechanics that in theory cannot be exactly fulfilled. Thus there is always some error involved, which varies from one case to another, and its significance will now be briefly considered.

To start with it should be noted that any difference in the value of Poisson's ratio will significantly affect the determination of the stress resultants only if the ratio of normal-stress deformation to shear deformation is an important factor in the elastic behaviour of the structural system concerned.

Thus the shear deformation in slender bar-type structures subjected only to bending and normal forces is so small that it is usually neglected even in theoretical analyses. For this reason the results of model tests on structures of this type can be applied to the prototype without any worry about the effect of Poisson's ratio.

In the case of bar-type structures whose members are subject to considerable torsional loads, the similarity condition $\mu' = \mu$ can be replaced by:

$$\pi = \frac{EJ}{GJ_T} = \frac{J}{(1 + \mu)J_T}$$

or

$$\left(\frac{J}{J_T}\right)^{1'} = \frac{1 + \mu'}{1 + \mu}\left(\frac{J}{J_T}\right). \tag{2.29}$$

From this expression, it appears that the possible error arising from differences in the magnitude of Poisson's ratio for the model and the prototype is not at all directly proportional to the two respective values of this ratio. Thus, for example, if $\mu = 0.2$ and $\mu' = 0.3$, the difference between $1 + \mu$ and $1 + \mu'$ is only about 7%.

In addition to torsion, if a beam is subjected solely to bending in one direction, it is always possible to satisfy condition (2.29) exactly by choosing a suitable cross-section in the model.

In plate- and shell-type structures, the interaction of shear force and normal force cannot be simulated by an adaptation of the geometry of the model. In such structures the effect of Poisson's ratio must show itself in the results. It is quite possible that stress resultants whose origin is largely attributable to μ (e.g. restrained transverse curvature) will, if a different value of Poisson's ratio is adopted, even change their algebraic sign. This phenomenon cannot, however, occur for quantities of the largest (absolute) magnitude, i.e. the most significant quantities that the investigator wishes to determine. Yet there is some uncertainty in the magnitude itself, and in order to cope with this it has been proposed that the actual stress resultant should be bracketed between two limits, one of which is determined with the correct μ' of the model while the other is determined from the relation

$$\bar{m}_x = \frac{1 - \mu'\mu}{1 - \mu^2}m_x - \frac{\mu' - \mu}{1 - \mu^2}m_y$$

where μ is the value of Poisson's ratio for the prototype. Now if the structure is designed for m_y and $\bar{m}_x$, the actual value of the stress resultant will have been catered for, or so it is assumed. Yet this is hardly a tenable assertion. The above 'conversion' of the stress resultants presupposes that the deflected surfaces

of the two structures are affine, despite the fact that they have different values of Poisson's ratio. This, of course, is not so. There is in fact no evidence that the unknown value is indeed bracketed in the manner claimed. In addition the design of a structure on the basis of two limit values is generally unacceptable on economic grounds.

Elastic model testing using materials which are very different from those used in the prototype is very common practice in connection with the design of reinforced concrete or pre-stressed concrete structures. The design procedures normally employed for such structures assume that within the context of the particular properties it is justifiable to adopt the stress resultants determined in the model—with its own particular value for Poisson's ratio —as a direct basis for designing the actual structure, i.e. not applying any adjustment. This can be justified as follows. In reinforced concrete design a value of $\mu = \frac{1}{6} = 0.1667$ is frequently adopted in the design and analysis of structures. This is really more a conventional rather than an actual value for Poisson's ratio for concrete. More recent research has shown that this value can be higher, i.e. in the range of 0.25–0.27. Hence conventional design procedures cheerfully accept a larger error associated with Poisson's ratio than the error committed in taking the stress resultants determined in, say, an aluminium model and applying them direct to the concrete structure.

Structures of reinforced concrete must, first and foremost, achieve the desired safety against failure. Irrespective of the effect of Poisson's ratio, the pattern of forces determined from measurements on a model always refers to a completely equilibrated system, as will also be true in the actual structure (prototype). Even if the latter is under-designed in certain members of secondary importance, it will in the event of overloading revert to a similar pattern of forces and state of equilibrium as a result of cracking of the concrete. Safety against failure is thus fully preserved. Slight deviations of the actual stresses from those determined in the model are quite acceptable under working load conditions.

In this connection it should be pointed out that if the finite element method of analysis with the aid of a computer is applied, generally larger (and unverifiable) errors are tacitly accepted, these being due to the approximation implicit in the assumptions for the state of stress in the interior of the structure or structural member concerned.

3

EXPERIMENTAL TECHNIQUE

3.1 MODEL MAKING

3.1.1 ELASTIC MATERIALS

3.1.1.1 Criteria for choosing the material

Precise knowledge of the properties of the material and their effects on the quality of the test results is a prerequisite condition for the success of the model test. All commercially available materials potentially suitable for making structural models offer technical advantages and limitations which must be weighed against one another. The ideal material does not exist, but efforts are continually being made to develop materials especially suitable for model testing.

The materials used for elastic-model tests must conform as closely as possible to the hypotheses on which general elastic theory is based. The required properties are: homogeneity, usually isotropy, a linear stress–strain relationship (Hooke's law) and an invariant Poisson's ratio. Materials possessing these fundamental properties within a technically useful range of stress or strain are, fortunately, abundantly available.

From the viewpoint of experimental technique, there are a number of further criteria to be considered, and it is these that make the choice of the model material somewhat of a problem. In this chapter we shall compare the available materials from this angle, i.e. the importance of individual criteria as far as experimental technique is concerned.

(i) *Modulus of elasticity:* A high modulus of elasticity requires the application of large loading forces in order to obtain reliable measured strains in the model. It also requires rigid bearings and complicated measuring arrangements unless the model is constructed to a very small scale. This makes greatly increased demands upon the precision of the model and inevitably involves a loss of accuracy in the measurements. On the other hand, if the modulus of elasticity is too low the stiffness of the strain gauges themselves can have a disturbing effect on the test results (Section 3.2.1).

(ii) *Poisson's ratio:* The error effect due to differences in Poisson's ratio has already been dealt with (Section 2.3.5).

(iii) *Creep:* All synthetically-manufactured materials undergo 'creep', i.e. deformation which is a function of time, temperature and stress. Whether this effect can be compensated by some artifice so that the results of the measurements can nevertheless be interpreted as quasi-elastic will depend on the outcome of a careful examination of these functional relationships. Such considerations will be presented later (Section 3.1.2.4).

(iv) *Thermal conductivity:* The conductivity coefficient is of importance in cases where strains have to be measured by means of bonded gauges. The lower the thermal conductivity of the material the greater will be the error effects due to heating up of the strain gauges (Section 3.2.2.2).

(v) *Workability:* The mechanical workability, or ease of machining, of the model material is very important from the viewpoint of the economy of the model test. With complex and geometrically irregular structure shapes, the choice of model material will be determined by the criteria of fabrication technology. Since it is, generally speaking, not ideally possible to combine favourable measuring and favourable fabrication properties in the same material, the choice will often have to be a compromise between these two sets of properties. If he has a well-equipped model-making workshop in terms of machine tools and other facilities, the model maker enjoys a greater freedom in the choice of material and can take full account of the dominant properties affecting measuring technique. A well-equipped workshop can be of major importance in the reliability and success of model tests.

The material constants for the different materials will not be listed here. Such data are to found in detail in the literature but as a rule such data should not be used for the evaluation of one's own tests. The important constants of the material employed must be determined by means of calibrating measurements for each set of experiments.

3.1.1.2   Metals

If exacting requirements are applied to the accuracy of the measurements, the most suitable materials are metals. Their behaviour generally conforms to the hypotheses of elastic theory. Among the metals, the aluminium alloys occupy a position of particular importance as model materials. They are characterised by large permissible strain values and high thermal conductivity in combination with a relatively low modulus of elasticity. Nevertheless, this modulus is still high enough—in contrast in particular with plastics—to minimise the disturbing influence of the inherent stiffness of the strain gauges. Poisson's ratio for these metals (*ca.* 0.30) is much closer to that of concrete than is Poisson's ratio for plastics.

The fact that metal, despite these ideal properties, is rarely used as a model material is mainly because of (a) the very high cost of machining and fabrication for complex models, (b) the high demands made upon the rigidity and precision of the sub-structure and (c) the heavy loading appliances required to produce the large strains necessary to give reliable measurements.

Figure 3.1 Aluminium model of a bridge deck, cut from the solid metal (Author's laboratory).

Because the material is difficult to work, there are almost insurmountable limits to the shaping of metal models. It is virtually impossible to construct 'freely' shaped shell structures in metal. Against this, metal offers such great advantages from the point of view of experimental technique that it is worth investigating the possibilities for further application. Part of the extra cost of making the models will in any case be offset by speedier and more reliable measurements. Better machine tools and the development of new methods of model making could secure a wider field for the metal model.

For solving simple slab or plate problems, metal is certainly the most suitable material. For one thing, to make a metal plate of any plan shape is simple; furthermore, for true slabs it is possible, in the model, to reduce the relative thickness of the material (Section 2.3.3.1) so that the requirements as to the sub-structure and loading arrangements are not so stringent. However, in the case of structures with variable thickness such as bridge decks, the reduction of the vertical scale must be made with caution (Section 2.3.3.3).

3.1.1.3   Plastics

A large number of plastics are available, most of which have at one time or another been used for model testing. Formerly, almost the only synthetic material of this type used for structural models was celluloid, but nowadays thermosetting plastics are extensively used, such as epoxides (e.g. Araldite) and polyesters, and thermoplastics such as polyvinyl chloride (PVC), polyethylene, acrylic resins, etc. From the mechanical point of view, all these plastics possess closely related properties. The modulus of elasticity is much lower than that of

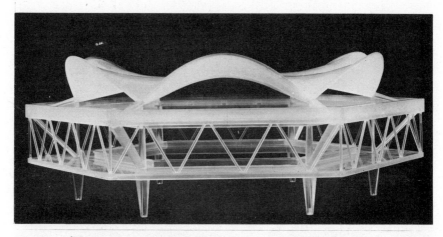

Figure 3.2   Model made of Plexiglass (Author's laboratory).

metals ($ca.$ $\frac{1}{10}$ to $\frac{1}{50}$ of the value), while Poisson's ratio is between 0.35 and 0.50, i.e. higher than that of metals. Plastics have low thermal conductivity which causes additional difficulties in connection with the use of electric resistance strain gauges, but they have the advantage of being relatively easily workable.

Among the plastics, acrylic resin (e.g., Plexiglass, Perspex) is pre-eminent for its advantageous properties. It is, accordingly, very popular and extensively used in model-testing laboratories. For one thing, acrylic resin is transparent, so that the relative positioning of two strain gauges on opposite faces of the material can be accurately verified. It can be bonded by an adhesive to produce completely monolithic joints. Any defects in such joints can at once be detected because of the transparency of the material. (Badly executed joints and connections can give rise to a totally wrong interpretation of the results of the measurements!) Acrylic resin is readily available in a variety of forms as semi-finished products and is easily workable either by various cutting techniques or by hot deformation, thus enabling it to be shaped and formed as required.

Although some of its technical properties are less favourable for model testing and measurement than those of certain other plastics (especially as regards creep), the properties of acrylic resin which are described in the following pages can be regarded as valid for all plastics. In specific cases, of course, appropriate coefficients should be determined experimentally.

3.1.1.4   Creep of plastics

If a plastic model on restraint-free supports is subject to loading by a set of constant forces which do not themselves undergo any change as a consequence of the deformation of the model, it will be found that the deformation that initially develops when the loading is applied will continue to increase

with time. This phenomenon is called 'creep' by analogy with the well-known phenomenon found in concrete under constant load, although the actual physical processes involved are very different in the two materials. The creep of concrete is only a partially reversible phenomenon, whereas in plastics the creep remains recoverable even after final polymerisation.

Accurate knowledge of the creep properties of the plastic used for constructing a model is of great importance to the correct interpretation of the measured results. The general laws and relationships that govern this phenomenon will therefore be examined in some detail.

The creep curves presented in Figures 3.3 and 3.4 were obtained by means of an electrical resistance strain gauge fixed to a calibration bar made of Plexiglass and were recorded by means of an X–Y recorder.

On considering any curve in Figure 3.3, we see that on application of the load an 'instantaneous' strain is produced which, while the load is sustained, steadily increases with time in accordance with a function $f(t)$. As appears from the diagrams, the rate of increase (i.e. the first derivative) of this function steadily diminishes however. These characteristics are typical of all plastics. Although the derivative $f'(t)$ continues to decrease, it never becomes zero.

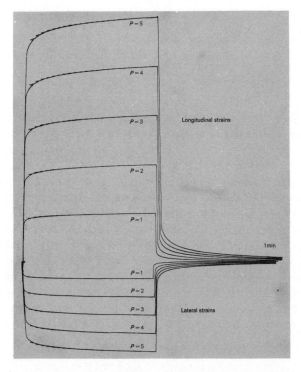

Figure 3.3 Creep curves for multiple loading determined on a Plexiglass bar. Each curve was obtained for sustained load followed by load removal and complete creep recovery. Upper curves: longitudinal strains. Lower curves: lateral strains. The invariance of Poisson's ratio emerges distinctly.

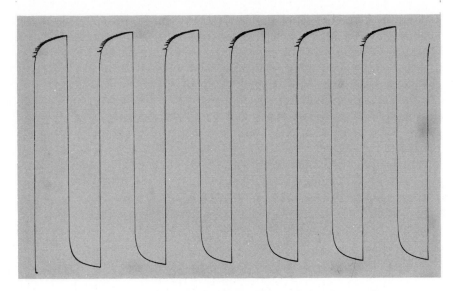

Figure 3.4   Loading and unloading cycles applied to the same bar.

This means that any representation of $f(t)$ by a logarithmic function, as is often encountered in the literature, must be regarded as no more than a fairly rough approximation. On removal of the load, the strain attained at time t, i.e.

$$\varepsilon = \varepsilon_0[1 + f(t)]$$

instantaneously decreases by an amount ε_0 and then creeps back (but now asymptotically) to the initial position. The creep recovery function on unloading is therefore *not* accurately mirror-symmetric with regard to the creep function. However, it is found that the recovery curve can, within the limits of accuracy of measurement, be made to coincide with the creep curve if the test specimen has been subjected to long-term sustained loading before removal of the load. This fact leads to the formulation of a hypothesis which also accounts for the asymptotic decrease of ε on unloading and certain other phenomena which will be considered later. It is assumed that the creep curve actually observed in any particular case is in part pre-determined by the 'loading history' of the material, i.e. that at any particular instant the creep curves corresponding to any previous loading undergone by the material are superimposed upon each other even though the corresponding stresses have long since ceased to act. According to this hypothesis, the creep recovery function can in this case be written as follows:

$$\varepsilon_e = \varepsilon_0[1 + f(t)] - \varepsilon_0[1 + f(t - t_b)]$$
$$= \varepsilon_0[f(t) - f(t - t_b)].$$

(Subscripts: e for unloading; b for loading)

It is therefore quite reasonable to find that ε_e tends to zero with increasing time, since:

$$\lim_{t \to \infty} f(t) = f(t - t_b).$$

In writing down the time-dependent function for ε, it is implicitly presupposed that as a result of assuming a dimensionless function for creep there is an affinity between creep curves for different values of ε_0. This assumption is indeed very accurate, as appears from Figure 3.3.

In order to determine the relation between ε_0 and the external loading, the test beam was subjected to a load which was increased in a stepwise manner. Between successive loading cycles, sufficient time was allowed in order to let ε_e creep back practically to zero and thus obtain a measurement not in-fluenced by the previous history of the specimen. It was found that the strains measured at equal time periods after the application of the load were propor-tional to the magnitude of the load applied. The strain function is therefore 'stress proportional', though it must be borne in mind that creep functions will also develop in the stress-free direction and must indeed so develop if the results obtained are to be interpreted in terms of elastic behaviour.

This strain behaviour can be investigated by means of the following ex-periment. On the same test beam, the strain transverse to the stress direction is measured for the same loading cases as in the previous experiment. Here again the strains are found to be proportional to the load, and the functions $f(t)$ are identical with those found in the stress direction. This means that Poisson's ratio actually remains invariant for all states of stress. In general, the strain in the plastic obeys the following relation:

$$E_0 \varepsilon_1 = F(t)(\sigma_1 - \mu\sigma_2)$$

where $E_0 = \sigma/\varepsilon_0$ for uni-axial stress in the direction ε_0. In the case of 'pure' loading of a specimen which has no 'previous history' we have:

$$F(t) = \frac{1}{1 + f(t)}.$$

If the correction function $F(t)$ and the instant of measurement are known, the result of the measurement can therefore be re-interpreted for a non-creeping material.

The foregoing principles may be summed up as follows:

Apart from a time function, ideal behaviour in accordance with the hypotheses of the elastic theory—with a linear stress–strain diagram and a constant Poisson's ratio—is observed with plastics suitable for making structural models. The time function for the modulus of elasticity may be reproduced but it is not a logarithmic function. The behaviour pattern for each individual case must be determined experimentally.

With automation of measuring procedures, the cyclic scanning of data from measuring gauges is becoming increasingly important. The loading history of the test specimen may significantly affect these results. It is therefore

important to make an analytical investigation of such cyclic scanning of data from a series of measuring positions under constant loading. Figure 3.5 indicates the time-dependent behaviour of the strain in a model which is subjected to repeated alternate loading and unloading during constant time intervals t_b and t_e, respectively. From the hypothesis for creep presented above, we can now write the time function for ε for two loading and unloading cycles (N denotes the number of full cycles):

$$\varepsilon_b^1 \Big|_0^{t_b} = \varepsilon_0[1 + f(t)]$$

$N = 1$

$$\varepsilon_e^1 \Big|_{t_b}^{t_b + t_e} = \varepsilon_0[f(t) - f(t - t_b)]$$

$$\varepsilon_b^2 \Big|_{t_b + t_e}^{(t_b + t_e) + t_b} = \varepsilon_0\{1 + f(t) - f(t - t_b) + f[t - 1(t_b - t_e)]\}$$

$N = 2$

$$\varepsilon_e^2 \Big|_{(t_b + t_e) + t_b}^{2(t_b + t_e)} = \varepsilon_0\{f(t) - f(t - t_b) + f[t - 1(t_b + t_e)] - f$$

$$- f[t - 1(t_b + t_e) - t_b]\}$$

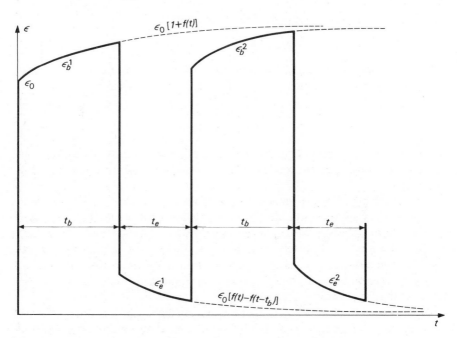

Figure 3.5   Explanation of how the creep curves develop under the cyclic loading.

or in general:

$$\varepsilon_b^N \bigg|_{t=(N-1)(t_b-t_e)}^{t=(N-1)(t_b+t_e)+t_b} = \varepsilon_0\{1 + f[t - (N-1)(t_b + t_e)]$$

$$+ \sum_{n=2}^{n=N} (f[t - (n-2)(t_b + t_e)] - f[t - (n-2)$$

$$\times (t_b + t_e) - t_b])\} \tag{3.1}$$

$$\varepsilon_e^N \bigg|_{t=(N-1)(t_b+t_e)+t_b}^{t=N(t_b+t_e)} = \varepsilon_0 \sum_{n=0}^{n=N-1} (f[t - n(t_b + t_e)] - f[t - n(t_b + t_e) - t_b]). \tag{3.2}$$

Thus the 'creep history' of a plastic specimen which has been loaded N times during constant time intervals to the same stress has been expressed in an analytical form.

We shall now calculate the difference $\varepsilon_b - \varepsilon_e$ at a particular instant t in the Nth cycle.

The sum in equation (3.1) can also be written as follows:

$$\sum_{n=2}^{n=N} = \sum_{n=0}^{n=N-2} (f[t - n(t_b + t_e)] - f[t - n(t_b + t_e) - t_b])$$

and thus acquires the same form as the sum in (3.2). The difference between the sums is:

$$\sum_{n=0}^{n=N-2} (1) - \sum_{n=0}^{n=N-2} (2) = -f[t - (N-1)(t_b + t_e)]$$

$$+ f[t - (N-1)(t_b + t_e) - t_b]$$

and hence:

$$\varepsilon_b^N - \varepsilon_e^N = \varepsilon_0\{1 + f[t - (N-1)(t_b + t_e) - t_b]\}$$

and since $t - (N-1)(t_b + t_e)$ always denotes the same initial period of time, independently of the cycle, we have:

$$\varepsilon_b(t) - \varepsilon_e(t) = \varepsilon_b^N(t) - \varepsilon_e^N(t).$$

This means that the difference functions are identical in each cycle. This result is important because the function ε_e is utilised as the zero reading (Section 3.3.3). Measuring cycles therefore remain fully reproducible.

The problem of correct strain interpretation becomes more difficult if the previous history of the creep of the test specimen has not been caused by cyclic loading always involving the same value of the stress. This problem is encountered with the plotting of influence functions associated with loads that are not stationary but which move about on the model. For instance, the algebraic sign of the stress measured at a particular point on the model may change between one loading cycle and the next. In such cases the creep history is liable to introduce a serious error into the measured values. Here too the computer can render useful service by storing the creep history of that point and using these data to correct the value actually measured.

3.1.1.5   Modifying the properties

The effect of fillers incorporated into polyester resin and epoxide resin (in particular Araldite) has been investigated in the Author's laboratory. The object of the experiments was to produce a material which, while retaining the good workability of plastics, would be more suitable for model-testing purposes than the commercially available materials.

The best results were obtained with Araldite filled with aluminium powder The following results were obtained with a mixture comprising 65 parts of aluminium to 100 parts of epoxide resin (parts by weight): (a) the modulus of elasticity was increased from about $32\,000$ kg cm^{-2} to $103\,000$ kg cm^{-2}; (b) Poisson's ratio was reduced from 0.42 to 0.32; (c) the creep function was reduced to about 10% of the original value; and (d) the thermal conductivity was very considerably increased.

In its measuring properties this material comes very close to aluminium, but is easily machined. Moreover it is suitable for bonding by adhesives. Theoretically, this material possesses excellent properties for structural models, yet it has not come into use in actual practice because of difficulties in processing it into completely homogeneous mixtures. If the epoxide resin used in preparing the mixture is too fluid, the aluminium settles too quickly; on the other hand, if the resin is too viscous it becomes very difficult to prevent air bubbles from forming.

3.1.1.6   Gypsum plaster

Gypsum plaster is widely used as a material for the construction of elastic models. The advantages of this material lie in its easy workability, its cheapness for large models, its Poisson's ratio (which is very close to that of concrete) and the possibility of controlling its modulus of elasticity by admixture of other materials and choice of water content. For these reasons, plaster is much favoured as a material for making models of dams, including the geological sub-soil of such structures. The use of plaster models for the simulation of reinforced concrete will be referred to later (Section 3.1.2.2).

The principal disadvantages associated with the use of plaster as a model material are its low tensile strength and the great difficulty of obtaining elastic properties which are uniform and accurately known.

The Laboratorio Nacional de Enghenaria Civil (LNEC) at Lisbon has tested hundreds of plaster models and thus probably has more experience with this material than any other institution. In that laboratory, a mixture of gypsum (plaster of Paris), kieselguhr and water is employed. The mix proportions of gypsum to kieselguhr is normally taken as 2:1. With this mixture, in conjunction with varying water-to-gypsum ratio from 0.8 to 3.0, a model material can be produced whose modulus of elasticity is adjustable to any desired value between $40\,000$ and 4000 kg cm^{-2}.

Originally the models were produced by casting the plaster in accurately dimensioned moulds. To obviate the formation of air bubbles, kieselguhr is pre-mixed with water for some hours before gypsum is added. To retard the setting of the gypsum, ice water is used.

It subsequently emerged that models made by casting have elastic properties at the surface which differ from those in the interior of the model, this difference being due to moisture movements in the material. For this reason, the models are now manufactured by machining (cutting and milling) from solid blocks which have been allowed to set and harden in air-conditioned rooms, at 40% relative humidity and 35°C, for at least one month.

Control of the stiffness (modulus of elasticity) of the material by varying the water content of the mix enables plaster models to be used for the simulation of the elastic interaction of different structural materials. In this way, the structural behaviour of a dam on foundation soil of varying quality can be

Figure 3.6   Plaster model of a dam (LNEC, Lisbon).

Figure 3.7   Making a plaster model (LNEC, Lisbon).

realistically simulated. Cracks in the sub-soil and even the effects of geological rock displacements can be taken into account. Of course, for making such composite models it is not possible to use only blocks of material that have been made and cured in the air-conditioned room under careful supervision. It is therefore not possible to reproduce the relative elastic properties of the various soil strata accurately in the model.

Nowadays efforts to simulate sub-soil stiffness properties by means of special mixtures of material in the model have been largely abandoned. Instead, the rigidity is controlled by drilling series of holes, with a greater or lesser spacing density, into the model. The effect of such holes is accurately known and in this way the modulus of elasticity can be varied up to a ratio of about 1:10.

3.1.2 MODELS MADE OF ACTUAL CONSTRUCTION MATERIALS

3.1.2.1 General

As has already been explained (Section 1.2), models made of the same materials as their prototypes must realistically reproduce the true behaviour of the latter under overloading and up to structural failure. Full-size tests can be carried out on structures composed of all practical materials and their combinations. Some of the most familiar are steel, timber, concrete (reinforced or pre-stressed), brick and glass-fibre-reinforced plastics. The use of these materials for making models involves difficulties arising from scale effects. As in the extreme case of the water drop (Section 2.1.5.4), these difficulties are, generally speaking, linked to the unavoidable fact that the surface area of geometrically similar objects does not vary in proportion to their volume and that the perimeter of a section through a structural member does not vary in proportion to the area of the section. As in a fluid, there are strength properties which are more particularly associated with the material in the interior of the model, whereas other strength properties are determined by conditions at the surface. The latter are usually considered to be negligible within the context of normal engineering design methods. In small models, however, their influence can be quite considerable. The strength (breaking stress) of steel wires is found to be greater with decreasing diameter (for the same grade of steel). Rather surprisingly, this trend is also found in wooden rods. Little research has been done on these phenomena and they cannot be explained in general terms. The model-testing engineer must therefore investigate his material experimentally in each particular case and try to obtain properties which make it possible to convert model-test results to values reflecting the behaviour of the prototype.

3.1.2.2 Problems of micro-concrete

Special importance is attached to the development of model materials which realistically simulate the mechanical properties of structural concrete,

which is now the most widely used and yet in some respects still one of the least understood construction materials. It may be used as plain concrete but more particularly as reinforced or as pre-stressed concrete. It is also frequently used in combination with structural steelwork (composite construction). The properties of concrete vary widely, depending on the quality of the aggregates, the water content and other aspects of the mix composition, the methods of mixing, placing and curing the concrete and other factors. Large sums of money are spent each year in research laboratories around the world on the investigation of the structural behaviour of this material. Yet the knowledge thus gained is comparatively meagre in relation to this effort

Figure 3.8 Model made of pre-stressed micro-concrete tested to failure (Laboratorio Central, Madrid).

and expense. In this situation the purpose of model testing is to extend the range of practical knowledge of what the structural engineer can do with concrete. For this type of work, a material known as micro-concrete has been developed possessing scaled-down properties of ordinary structural concrete.

Successful, but as yet relatively few, model tests with micro-concrete have been performed in research laboratories in various parts of the world over a good many years. The main reason for this relative neglect of the material is that the scientific basis has not yet been firmly established. At the present time, Committee 444 of the American Concrete Institute under the chairmanship of Professor R. White is engaged in research and development work with the aim of providing such a basis. Pending the results of the Committee's work, the problems and possibilities associated with the use of micro-concrete will not be further considered here.

3.2 EXPERIMENTAL PROCEDURE AND EQUIPMENT

3.2.1 GENERAL PRINCIPLES

The engineer is usually pleased if measurements performed on a completed structure prove to be within $\pm 10\%$ of advance calculations. Bearing reactions will generally be somewhat more accurate. On the other hand, local stresses may frequently differ by many times more than 10% from the predicted values. From this fact (for which allowance is made by the introduction of safety factors in structural design calculations), it is often concluded that there is no point in striving for a high degree of accuracy in model measurements. This is a fallacy, not because the prototype structure will necessarily behave exactly as the model, but because accurate measurements are essential for a correct interpretation of the results. For calculating stress resultants in the model, certain analytical operations are performed upon the measured values, and these operations may magnify the error in the model in relation to the prototype. Thus, membrane stresses are determined as differences of extreme fibre stresses. Check calculations for equilibrium, too, are reliable only with precise measured values. The precision of the measurements will depend upon the quality of the electrical measuring technique (Section 3.3) and upon the properties of the test apparatus.

The test set-up comprises the bearing system for supporting the model, the base which absorbs the bearing reactions and the loading equipment. Each of these elements must satisfy the basic conditions for the test within known limits of tolerance.

The requirements applicable to the base, or sub-structure, can be formulated quite simply. The base frame and its support on the ground must be so rigid and strong that its deformations must not exceed about 1% of those expected to occur at the edges of the model. An approximate comparison of the rigidity of the model with that of the steel beams of which the base is constructed will usually suffice for the purpose.

The boundary conditions of the model are simulated in its bearing system, which forms the connection between the model and the base. Universal bearing elements with which all boundary conditions (rotational fredom, or elastic restraint as regards rotation and deflection, etc.) can be simulated in

a simple way are not commercially available. Some special devices will be described in the next section.

The measurement of external reactions is of great importance for a correct understanding of the structural behaviour of the model and prototype. Because of the technical difficulties associated with their measurement, experimental investigators have often tended to neglect the precise determination of bearing reactions. As a general principle, all external reactions, whether these be forces or moments acting in any direction, should always be completely monitored. The accuracy of the experimentally determined values can (in contrast with strain measurements) be reliably verified by means of equilibrium checks, comparing the measured reactions with the known external loads applied to the model. If such checks confirm the trustworthiness of the reaction measurements, it is possible, in combination with the external loads, to calculate the total internal forces at any particular section of the model within a known tolerance (Section 2.3.1).

Strain measurements give information only on local stresses. There is no guarantee that the measuring positions chosen include the parts of the structure which are subject to extreme stresses. We should therefore install a series of very closely spaced gauges at a selected section of the model. The results of the strain measurements will provide a check on the accuracy of the assumed material constants and enable the stress resultants to be compared with the very reliable values calculated from the reactions. This is the only way to make a dependable assessment of the accuracy of measurement actually achieved.

Figure 3.9 Model for the new municipal theatre at Basle (Author's laboratory). Design: F. Schwarz, R. Gutmann, U. Gloor, Architects; H. Hossdorf, Engineer.

Figure 3.10 Consistently applied three-dimensional bearings for a model providing a support system whereby the results of the tests can be reliably checked with the aid of equilibrium calculations.

It also checks that the stress distribution occurring at that section can be satisfactorily detected with the arrangement adopted for the gauges.

3.2.2 BEARING SYSTEMS FOR MODELS

It has already been mentioned that the simulation of general bearing conditions runs into considerable technical difficulties. Some of the more recent developments will be described here with reference to examples.

The great obstacle to all efforts to satisfy the required boundary conditions is friction. It is often necessary to provide freedom of movement in one direction or another at the bearings. Frictional effects are often the underlying cause of inexplicable results of measurements (and this applies also to measurements performed on actual completed structures). The practical difficulties of constructing frictionless bearings are often not appreciated. The magnitude of the frictional forces can never be determined and it may even change from positive to negative. Because of their inelastic and sometimes random character, frictional effects can be identified when the results of measurements with identical loads, applied in different sequences, are not reproducible. The reasons for this will be explained with the aid of a simple example.

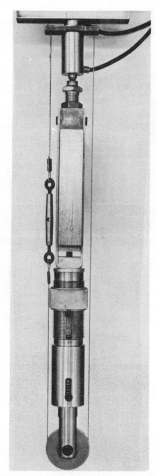

Figure 3.11 Pre-load applied to rocker-type model bearing and measuring transducer by means of tensionable springs. The pulley wheel permits frictionless rotation of the model.

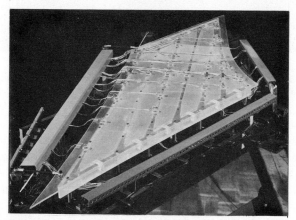

Figure 3.12 Curved and skew box-girder bridge deck on elastically deformable bearings to simulate the flexibility of neoprene (synthetic rubber bonded) bearings (Author's laboratory; design by the firm of Polensky & Zöllner).

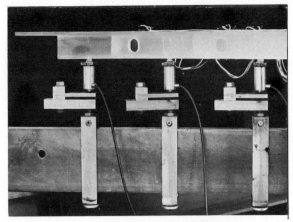

Figure 3.13   Close-up view of bearings.

Figure 3.14   Edge support of a triangular slab on a continuous girder with variable stiffness (Author's laboratory).

Suppose we wish to determine the influence line for strain in an extreme fibre at the centre of the right-hand span of a beam which is continuous over two spans. Assume that the freely-rotatable central bearing develops friction. If the load on the left-hand span moves from left to right, the moment occurring over the central bearing will at first be reduced by the friction so that the effect of the movement on the strain measurement will also be less. In moving along the left-hand span, the load will reach a point where the direction of rotation of the beam over the central bearing changes. When this happens the friction

will change its algebraic sign. From this point onwards the moment over the bearing will be increased, i.e. it will become too large. This change-over point corresponds to a break in the continuity of the influence line. If the influence lines are plotted for a load travelling from right to left, the effect of the error will also be reversed.

Special rocker bearings provided with spherical steel tips resting in bronze bearing cups have been developed for the measurement of bearing reactions. In contrast with the load cells commonly used for such measurements, these special bearings ensure freedom of rotation and displacement.

The devices used for measuring the reactions should also be able to measure tensile forces. As the measuring devices are not designed to take tensile loads, they are given a precompression which is kept constant during the testing of the model and which must be larger than the tensile stresses expected to occur. The pre-compression is achieved by applying a suitable pre-load to the model. By appropriate adjustment of the electrical measuring equipment, the pre-compression is considered to be equal to zero on the scale of measurement. The pre-load may take the form of actual weights applied to the model. However, this procedure is often objected to because it interferes with the application of the test loads. To overcome this difficulty, the pre-load may be applied by a system of springs whose tension can be adjusted.

Any desired degree of freedom may be obtained by combining a number of rocker bearings which maintain a rigid supporting element in three-dimensional equilibrium. From the known geometry of the measuring devices, it is possible to calculate the magnitude, direction and position of the resultant in space.

The above principles and techniques are illustrated in the accompanying photographs and explained further in the captions.

3.2.3 APPLICATION OF EXTERNAL LOADS

3.2.3.1 Introduction

Much imagination and effort have been devoted to the problems of devising suitable loading devices. This is hardly surprising if it is borne in mind that the speed and convenience with which a planned test sequence can be performed are largely dependent on the serviceability and adaptability of the available loading system. Hence the development of flexible equipment for rapid and methodical application of forces is of major importance in efforts to 'dematerialise' the model test. The ideal loading device has yet to be invented and no two model-testing laboratories use identical equipment.

In this section some of the better-known types of loading apparatus will be reviewed and their advantages and disadvantages critically examined with a view to obtaining some idea of the requirements that the ideal device would have to fulfil.

In attempting to make a systematic classification of the loading principles, it soon emerges that there are two main classes:

(i) Equipment for carrying out first-order elastic tests which applies *constant* loads or groups of loads to the model in conformity with the defined linear relation between loads and measured load effects (deflections, etc.). The choice of the magnitude of the loading will depend on a technical consideration of the measuring facilities and test set-up (good readings of the measured values, avoiding membrane effects in thin models, stability of the base, etc.). The effect of the loading acting on the prototype is calculated by determining a proportionality factor.

(ii) Equipment for carrying out non-linear elastic tests or loading tests to failure. The absolute magnitude of the range of loading to be applied can now no longer be freely chosen but is instead determined by the model laws. It must be possible to vary all the loads or groups of loads simultaneously and preferably in a continuous manner during testing. Equipment belonging to this category is designated as *continuous loading apparatus*.

Before dealing in more detail with the various technical appliances, some theoretical aspects of the application and positioning of the loads will be briefly discussed.

In many experiments the action of distributed loading will have to be approximated as accurately as possible by a number of individual concentrated loads. Two important considerations apply with regard to such cases:

(a) The load application points should, if possible, be remote from the measuring positions. Surface strains affected by the local three-dimensional state of stress at the load application point are liable to make the measured results worthless.

(b) The arrangement and number of individual loading points should be chosen so that the deformation at the measuring positions conforms as closely as possible to the deformation that a distributed loading would cause.

The criteria for the arrangement of the loads, which will be explained with reference to an example of a beam loaded in bending, are generally valid also for slab and shell structures.

The transducers (strain gauges, etc.) should, if possible, be installed in those positions on the model where extreme stress and strain conditions are expected to occur. They must certainly be located along edges, over bearings and in mid-span regions. Because of this requirement, the application of loads over the bearings and at or near the centres of the spans will have to be ruled out. Ideally, a group of loads for simulating a distributed line load or patch load will comprise an even number of individual (concentrated) loads, spaced equal distances apart, while the first and last loads on a span will be located at a distance from the bearing equal to about half the spacing of consecutive loads in the group.

This arrangement is illustrated in Figure 3.15, which shows that in the case of a simply-supported beam the mid-span bending moment will, irrespective of the number of loads in the group, have a value exactly corresponding to uniformly distributed loading. In statically indeterminate systems this is no

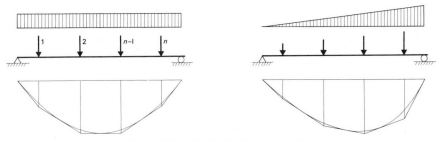

Figure 3.15 Statically determinate beams.

longer so, since here the strain energy associated with the loading becomes greater as the number of individual loads applies is smaller. The diagrams in Figure 3.16 give some idea of the magnitude of the error due to the application of concentrated loads instead of distributed loading in relation to the number of loads. It emerges that the error depends to a considerable extent on the type of structural system considered, and that if 10 concentrated loads are applied to simulate distributed loading the error involved in this approximation will not exceed 1 %.

3.2.3.2   Load arrangement

At first sight, it would appear desirable to be able to apply (distributed) patch loads as well as concentrated loads; indeed most conceivable systems for applying distributed loading have been tried at one time or another. A few of them may be briefly mentioned here:

(i) Heaping granular materials such as sand or shot (metal pellets) on the model. With this method the internal friction in such materials may cause the actual load distribution to differ significantly from the desired distribution.

(ii) Standing an open column of liquid on the model (in the case of plane models) which must therefore form part of the bottom of a tank or vessel holding the liquid. Sealing is effected by stretching a thin rubber membrane across the bottom. When the model undergoes deformation, unpredictable forces develop at the edge of the rubber at the joint.

(c) Uniformly distributed area loading produced by filling a closed rubber cushion with liquid or by inflating it with air.

All these methods are clumsy and usually inexact. They also limit the choice of other forms of loading that can be applied. In individual cases such methods may be advantageous for the investigation of special phenomena, using a loading installation designed to perform a specific function. Generally speaking, however, these methods have no place in modern model testing which aims to achieve a high degree of mobility and versatility.

Experience has shown that efficient apparatus can be constructed only for the application of concentrated loads. Distributed loads must therefore be simulated by groups of concentrated loads. The number of such loads needed should be kept as small as possible.

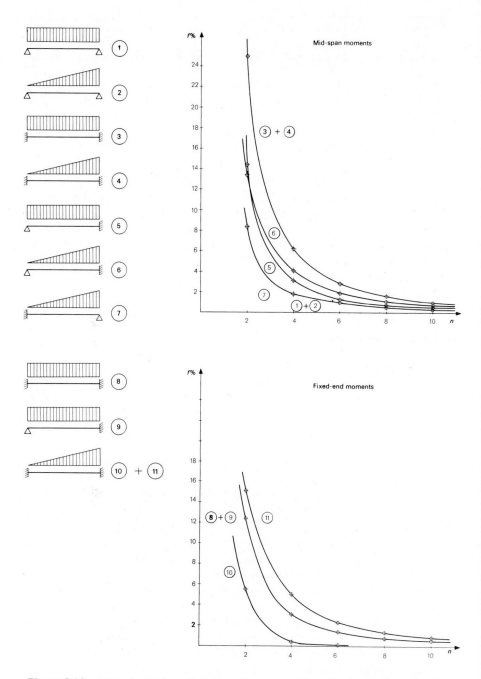

Figure 3.16   Errors due to the substitution of concentrated loading for distributed loading.

Figure 3.17   Simulation of hydrostatic pressure on a model dam by filling a rubber bag with mercury (LNEC, Lisbon).

Figure 3.18   Simulation of dead weight stresses by constructing the model in successive stages and superposition of the external loadings which act at each stage.

3.2.3.3   Linear loading devices

The number of measurements to be performed and recorded during a model test is equal to the product of the number of measuring positions and the number of loading cases to be investigated. The 'loading cases' are usually produced by the application of groups of loads comprising an appropriate number of concentrated loads.

In nearly all model-testing laboratories, it is still standard practice in each case to apply such load groups simultaneously in order to minimise the number of measurements to be performed. The inconvenience and time involved in correctly determining and rearranging the weights is preferred to the tedious and time-consuming tasks of recording and processing an excessive quantity of measured data. On the basis of this approach to the problem, various methods and appliances for load application have been evolved which have to some extent become standardised.

The weights applied to the models nowadays usually consist of slender steel tubes or rods suspended by means of wires or threads. The tubular weights can be adjusted by filling with, for example, lead shot (Figure 3.19). Although the work of preparing large numbers of such weights can be rationalised, the task of changing over from one loading case to another can nevertheless take hours.

Figure 3.19   Device for preparing and adjusting individual weights by filling them with lead shot (Author's laboratory).

A piece of equipment which has come into widespread use for supporting the weights under the model and for the controlled application of the loads to, and their removal from, it is the *lifting table*. The cylindrical or prismatic weights stand on a flat platform which can be raised and lowered vertically by electric, hydraulic or pneumatic power. When raised, the table top supports the weight and thus relieves the model of load; when it is lowered, the weights are suspended freely from the model. By means of the lifting table rapid loading and unloading of the model is possible. The timing of the loading cycle can be accurately controlled, this being particularly important for the elimination of creep effects in models made of plastics (Section 3.1.2.4).

In order to reduce the number of individual weights and thus cut down the amount of work involved in changing the loading conditions of a model, load-distributing lever systems are frequently employed as a means of producing large load groups. Some laboratories use controlled elastic forces instead of weights for obtaining the closest possible approximation to distributed loading (Fig. 3.20). In such an arrangement, the load application points on the model may be connected to the top of the lifting table by thin rubber cords, the magnitude of the loads now being determined by the distance the table is lowered (and by the deformation of the model itself).

The above loading methods and appliances are still in common use, but are not compatible with modern measuring techniques. The argument that

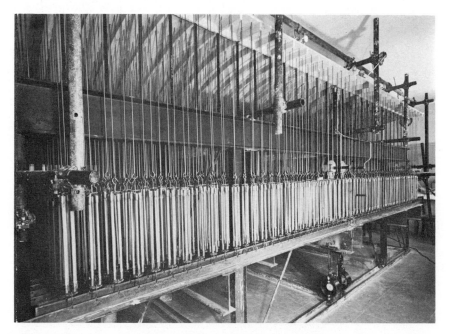

Figure 3.20 Applying a continuously increasing area loading to a model for testing to failure. The weights, attached to the load application points on the model by means of elastic cords, are supported by a steel table which is gradually lowered (Laboratorio Central, Madrid).

it is essential to reduce the number of measurements so as not to make their recording and evaluation unmanageable is really no longer valid. With modern electronic equipment it is possible to perform measurements accurately and reliably in a fraction of the time required by conventional manual methods only a few years ago. With up-to-date techniques the effects of individual loading cases can be found by superposition of the influences produced by a single load whose point of application is systematically changed on the model. In other words, we produce influence functions.

To illustrate the principle, let us consider the application of the method to the problem of determining the distribution of bending moment in a slab or plate under uniform loading. This is a very unfavourable example, and the procedure of tackling the problem by influence functions will, at first sight, appear quite far-fetched. For instance, to do the job properly it may be necessary to perform a hundred measurements for each of a hundred load application points, i.e. 10 000 measured data in all. With a loading device as illustrated in Figure 3.26, controlled by a program fed in on punched tape, the load is successfully applied automatically to the application points and the requisite hundred measurements is recorded. If the stream of information thus obtained is stored on magnetic tape, the whole operation of recording the 10 000 measured values takes about 25 min. The data are then processed in a computer and the result for superimposed uniformly distributed loading will be available in half an hour so that the testing can be completed much more quickly than by manual methods.

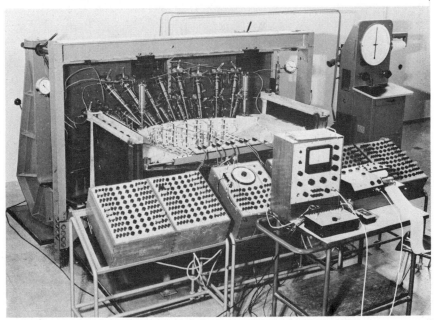

Figure 3.21   Universal hydraulic loading apparatus for the testing of model dams. A notable feature is the ease with which the individual loading jacks can be installed in any desired direction (LNEC, Lisbon).

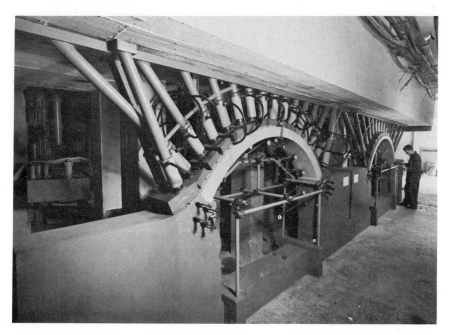

Figure 3.22   Hydraulic loading jacks mounted on a model arch rib (ISMES, Bergamo).

Figure 3.23   Horizontal loads produced by hydraulic jacks (ISMES, Bergamo).

Figure 3.24 Manually-guided tracer which, while the deflection of the pointer of the indicating instrument of a measuring bridge is kept constant, enables the contour lines of influence surfaces to be directly plotted (Author's laboratory).

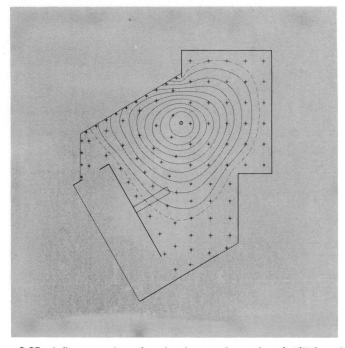

Figure 3.25 Influence surfaces for a bearing reaction under a freely shaped slab.

Figure 3.26   Punched-tape-controlled carriage-mounted device for automatic application of load at any desired point, determined by co-ordinates, on a structural model. The vertical force can be varied freely between 5 and 60 kg. The electronic control apparatus is illustrated in Figure 3.45 (Author's laboratory).

With automated and computerised determination of the influence values we have gained much more than the solution of a particular problem. The data storage system now contains the information needed for the computer to simulate all conceivable types of loading that may be applied to the plate under investigation. A 'loading case' (i.e. any particular combination of loads) can be analysed in a few minutes, including the input data for the pattern of loading.

Thus the model, functioning as an analog input device, has served its purpose and can then be scrapped. Yet, thanks to the stored data, new loading cases on that particular plate can be analysed long after the actual model has ceased to exist. This is very valuable because loading cases other than the one originally envisaged often have to be subsequently investigated.

Besides affording the advantage that the model need not be kept available for a long period (in case it is needed for further testing), the electronic storage of the influence functions also makes possible a clear separation of the work of the design engineer and the function of the model-testing laboratory. Magnetic tape containing stored information can be sent to the engineer however far away his place of work may be from the original testing laboratory, and he can process the data to suit his requirements.

Other aspects of this fundamental change in outlook with regard to model testing are considered in Chapter 4.

Nothing has yet been said about producing test loads acting in a particular direction. In most cases the loads applied to the model are gravity forces because the use of weights is convenient and accurate. However gravity has the disadvantage that it does not readily lend itself to acting other than in the vertical direction. Let us consider once again the use of the lifting table for applying and removing test loads. To employ this method for producing obliquely-directed (inclined) forces, the load must be applied by means of a string over pulleys. The construction of complex and frictionless pulley systems or similar mechanisms is more laborious than the relatively simple operation of suspending vertical loads from the model. If it were possible to construct a loading apparatus which would apply loading in any direction (in three dimensions) at any desired point in space, the problem of elastic-model testing would be solved in principle. An apparatus designed to fulfil this requirement is at present under construction.

3.2.3.4 Continuous loading devices

In this field of elastic-model testing, experimental techniques will soon attain a degree of perfection which is unlikely to be improved upon in any fundamental way in the foreseeable future.

The same cannot be said for the loading devices used for non-linear (inelastic) testing and for tests up to failure. It is in these sectors particularly that the need for rapid development appears most urgent because the computer can make only a small contribution towards solving the problems with which we are faced.

Loading devices available at the present time are still rudimentary. An ideal device should be able to apply forces in any desired direction in anything up to a hundred points (say) on the model, and each individual force should be controlled in accordance with its own program or be capable of individual adjustment on the basis of measured results processed on-line. As yet no such machine exists although there are no technical reasons why it should not be

constructed. The functions of the loading installation will have to be monitored by a suitable computer equipped with a rapid interrupting system.

So much for the near future. But even at an earlier period, before the vast possibilities of modern electronics became available, adequate loading devices for carrying out tests to failure were designed and used, although their versatility and adaptability left much to be desired.

Although it is no longer advantageous to construct such a piece of equipment, mention must nevertheless be made of the ingeniously simple continuous loading device constructed by E. Torroja. This device functioned on Archimedes' principle, the model being built over a tank of water about 2 m in depth. The loading units consisted of cylinders floating vertically in the water which were so adjusted for weight that they lay flush with the water surface. Each cylinder was connected by a slack thin string to the corresponding load application point on the model. When the level of the water in the tank was lowered, each cylinder exerted a vertical force through the tightened string, the magnitude of this force corresponding to the reduction in the water displacement of the cylinder.

This loading installation is very cheap and produces extremely accurate loads which can be very finely controlled. It has some disadvantages however, i.e. inconvenience of working over and in water, the slow rate of load change, the maximum attainable loads being too limited for many experimental purposes and the application of oblique (non-vertical) loads requiring much extra effort and elaboration of the equipment.

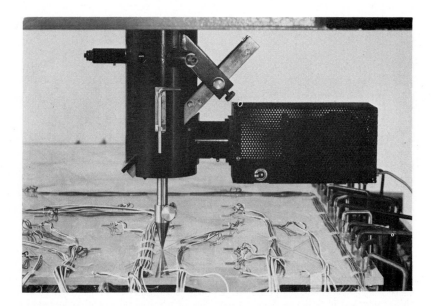

Figure 3.27 Detail of the replaceable load application pin. For programing the loading positions, the pin is removed and temporarily replaced by an optical projection device which forms an image of the crosswires on the model.

Figure 3.28 Device for the external application of prestressing forces to an elastic model of a weir (Author's laboratory).

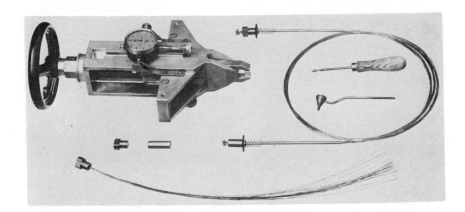

Figure 3.29 Equipment for producing realistic pre-stress in 1:20 to 1:5 scale models. It comprises a mechanical tensioning device with gauge giving a reading of the force applied; a 'cable' composed of 0.05 mm diameter high-tensile steel wires; a fixed anchorage head to which the ends of the wires are secured by embedding them in a zinc alloy (poured molten around them); spacer elements for securing the cable extension; and special tools for forming the anchorage of the wires (Author's system).

More recent devices usually operate with groups of hydraulic jacks. Failure tests on dams, which require particularly large loads, are performed exclusively with hydraulic applicances. The model-testing institute at Bergamo, Italy, has done valuable pioneering work in this field, and at the Institute at Lisbon a new and highly advanced loading device, built specially for the testing of dam models, is in use. This device is notable for the ease with which it can be adapted to a wide variety of shapes for dams.

Whether hydraulic loading systems have much of a future in model testing is very doubtful, however. The use of high-precision hydraulic jacks to meet the demand for a much larger number of load application points per model seems uneconomical. On account of the attendant risk of dirt and contamination, the use of complicated oil-operated hydraulic systems in the laboratory is hardly desirable. Electromechanically operated appliances equipped with load cells for controlling the magnitude of the forces generated would appear better suited for future installations.

3.2.4 PRE-STRESS

3.2.4.1 Two possibilities of simulation

Although the pre-stress in a structure is usually produced by means of tendons (cables, wires) embedded within the structural material, in effect pre-stressing is no different from the perhaps somewhat more complex method of external loading, at any rate so long as the structure remains within the elastic range of behaviour. Hence it must be possible at the working load of the structure to simulate the pre-stress by means of externally applied loads. However, if the structure is loaded far beyond the elastic range and up to failure, a non-linear mechanism will develop and further behaviour will then depend on the properties of the concrete and steel and on the bond between these two materials. Meaningful information on the behaviour of the structure at failure can therefore be obtained only from models in which these phenomena and properties can be adequately simulated. As for loading devices, a choice has to be made between the two possibilities for devices for producing pre-stress depending on the nature of the problem to be investigated.

There is another important reason why the loading due to pre-stress must be subject to separate consideration. At the design stage of the structure the engineer treats this loading case in a fundamentally different manner from the other external loads. Whereas live and dead loads are pre-determined by the relevant codes of practice or design standards and by the type of structure chosen, the optimum pre-stress must at this stage still be determined. It must therefore still be possible to vary both the magnitude and the geometric disposition of the pre-stress in the model test as well.

Figure 3.30 The pre-stressing apparatus illustrated in Figure 3.29 is shown here in use in a buckling test on a V-section beam. The test loading is applied by means of an ordinary tensile testing machine in combination with a special steel frame.

Figure 3.31 Test on a model, made of pre-stressed micro-concrete, for prefabricated northlight roof units (Author's laboratory).

3.2.4.2   Simulation by external loading

Even in elastic models, attempts are occasionally made to simulate the pre-stress realistically by means of tendons. This procedure is troublesome and clumsy, since the line of action of the pre-stressing force once adopted cannot subsequently be changed except by mechanical alterations to the model.

Particularly in conjunction with the influence function method, it is possible to rapidly simulate the action of any desired tendon profile without recourse to physical intervention in the model. This procedure, which provides the designer with a reliable and economical means of freely determining the pre-stressing tendons and their positions in the structure, is described in Section 4.2.3. It need only be noted that when the tendon profile has been geometrically determined and the initial pre-stressing force is known (with some appropriate allowance for friction), the load distribution on the model may be accurately determined. It can always be divided into three parts:

(i) A normal force, a shear force and a moment at the anchorage.

(ii) A linear load due to radially directed forces on the concave side of curved tendons.

(iii) In the case of post-tensioned cables in pre-formed ducts, a linear load in the direction of the tendon caused by friction in its sheath or duct.

For plane structures such as bridges, this complex pattern of loading due to pre-stress is often simplified and approximated in the model-testing procedure and evaluation. The effects of normal force are determined by calculation,

Figure 3.32   Pre-stressing system for models as used at the Laboratorio Central (LCEMC), Madrid.

and only the vertical components of the radial forces (due to tendon curvature) and friction forces are taken into account. These vertical forces are applied as external loading to the model.

In combination with hybrid analysis (Section 4.2), the procedure can be refined to a theoretically exact method. The application to the model of the loads required to produce radial forces can be dispensed with altogether, the computer simulating the loading by manipulation of influence matrices.

3.2.4.3   Failure tests and pre-stress

The pre-stressing details for any particular structure are usually designed in two stages. First, the optimum tendon profile and distribution of pre-stressing force are determined for working load conditions. This analysis is based on elastic theory or on the results of elastic-model tests, as referred to in the preceding section. Next, with the tendon dimensions obtained, the structure is analysed for ultimate load conditions in order to determine its safety against failure. In most cases, the structural safety can be reliably determined analytically by simple calculations. Hence the need to perform tests up to failure arises only in rather exceptional cases. The following provide some examples:

(a) The distribution of the pre-stressing force in the anchorage zone cannot be accurately determined by analysis. In particular, this occurs at the edges of pre-stressed concrete shell structures with cable anchorages at relatively wide spacing.

(b) It may occur that at failure the effective cross-sectional area of concrete, and therefore the internal lever arm between tensile and compressive zones, cannot be reliably estimated because the shear failure mechanism that develops cannot be correctly assessed. This, too, is a problem typically associated with shell construction.

(c) There may be stability problems arising during failure.

(d) Failure phenomena in thick-walled (three-dimensional) structures. These may occur in connection with pre-stressed concrete nuclear reactor vessels.

Special devices for the realistic simulation of pre-stress have been developed at various testing laboratories illustrated in the accompanying photographs and briefly described in the captions.

3.2.5   DISPLACEMENT OF SUPPORTS

Applying specific amounts of deflection or rotation to structural models is as important to experimental technique as the application of external loads. From the design engineer's point of view, the occurrence of movement at supports is just another loading case—often a very important one, which in

Figure 3.33   Electropneumatically-controlled displacement measuring transducer (Author's laboratory).

actual structures is liable to occur as a result of movements of the foundations.

Ideally it would be desirable to measure the influence of three unit translations and three unit rotations for each point of support. The effects of any particular movements that develop at the bearings of a structure could then be determined simply by superposition. However, ideal measuring instruments for the three-dimensional detection of displacements are not yet available, just as ideal three-dimensional bearing reaction measuring instruments are still in the future. Developments are in progress, however, as exemplified by the electropneumatically-controlled one-dimensional displacement detector (Fig. 3.33).

3.2.6   DEAD WEIGHT

Simulation of this loading case presents no major problems in structural and bridge models. In the case of elastic models the full dead weight (self-weight of the structure) is applied as external loading. But models to be tested to

failure (scale $l':l$) must, in addition to the dead weight, be subject to an area load $g(1 - l'/l)$. Theoretically this procedure is not exact since the effect of dead weight in the interior of the structure is neglected. In a concrete structure this internal effect corresponds to a pressure of $0.25 \, \text{kg cm}^{-2}$ for each metre of height; this pressure, or stress, is relatively so small that it can often be ignored.

This is not so with structures in which the internal stresses are a result of the action of the dead weight itself. High dams are an important example of this. The difficulty here is that for reasons bound up with similitude mechanics the stresses due to dead weight in the model are extremely small and no zero measurement can be performed without a known change of weight. As may be imagined, it is technically awkward and complicated to simulate dead weight by individual loads applied in the interior of the model. Other methods must therefore be sought.

Theoretically, the simplest solution is provided by the principle of *inversion*. The readings obtained from the various measuring stations on the model in its upright position are all adjusted to zero. Then, by means of a suitable rig, the model is rotated through 180° about a horizontal axis so that it now hangs from its bearings. The effect determined at the measuring stations will correspond to the stresses and strains which would occur if the model were loaded with twice its dead weight. In small models, however, this effect is hardly measurable, while for large models this method suffers from the drawback that the rig for rotating the model becomes elaborate and expensive.

If a large centrifuge is available, the dead-weight effect can, in suitable cases, be more readily determined by subjecting the model to a multiple of the earth's gravitational acceleration.

For producing dead-weight stresses in models of dams, the LNEC Laboratory, Lisbon, uses a method which is of theoretical interest and which produces measurable values five to 10 times larger than by the inversion method. It can be shown that the state of stress in a body suspended upside down in a liquid and loaded by the buoyancy force is proportional to its state of stress due to dead weight, provided that the appropriate boundary conditions are satisfied by the suspension system. If mercury is used as the immersion liquid, deformations of measurable magnitude can be produced in the model.

In recent years, the stresses in large dams have been experimentally investigated during the various stages of construction. Thus the model must also be constructed in successive stages. This provides the opportunity for externally loading it at each stage and measuring the stresses produced. In this way, the effect of dead weight can be determined by superposition for each intermediate stage in the completion of the dam.

3.3 ELECTRICAL TECHNIQUES OF STRESS ANALYSIS

3.3.1 INTRODUCTION

This book is concerned with model analysis and the testing of structural models and it would be outside its scope to give a comprehensive and detailed treatment of electrical methods of measurement for the determination of stresses, forces and deformations. The reader wishing to study these matters more thoroughly may usefully seek information from the many manufacturers of such measuring equipment.

In the present chapter, therefore, we must confine ourselves to taking a look at these techniques and instrumentation (which are of major importance in connection with the development of experimental methods) from the viewpoint of model testing. Actual electronic problems will be considered only where essential to a proper understanding of the principles concerned. More particularly, this discussion will attempt to indicate criteria for judging and assessing the merits of various methods with the aim of providing a guide to the many possibilities that electrical metrology and instrument technology can offer.

Mechanical measuring devices will not be discussed here at all. An understanding of their functioning is within the experience of any engineer. Besides, the large-scale use of mechanical equipment in this field of research is largely obsolete. It is clumsy and not very suitable for automation, certainly not in terms of the working speeds attainable with electronic computers. However, mechanical devices have the great advantage of being robust and reliable and for this reason they continue to occupy a secure place as calibrating instruments.

The future of model analysis lies in the utilisation of modern electrical measuring techniques in combination with the computer (Chapter 4). We shall accordingly classify these devices from this point of view in relation to the computer (Fig. 3.34).

To begin with, a distinction is to be made between the *digital* and the *analog* representation of physical or geometric values. Analog values (which may, for example, be viewed on a screen or by the deflection of the pointer of an indicating instrument) have the distinctive property that they give a continuous representation of a physical process. With electrical techniques the measured

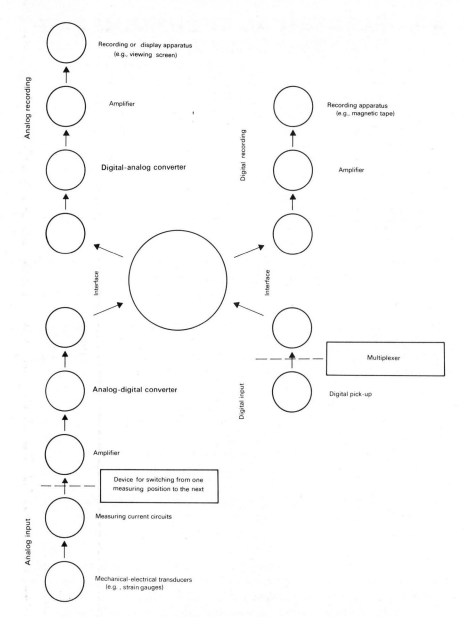

Figure 3.34   Data interchange between computer and peripheral equipment.

variable is generally represented as a voltage to which it is linked by a known, but not necessarily linear, relationship. On the other hand, digital representation is essentially numerical. It is discontinuous and only able to indicate *discrete* values in a more or less closely spaced sequence. The finer the division, i.e. the closer together the measured values are located in time, the more accurately will the digital representation conform to the actual continuous process.

Its great advantage consists in being able to present these values as numerical quantities and to perform arithmetical operations with them immediately.

Digital signals also have the advantage of lending themselves much more reliably to teletransmission. The signal level arriving at the receiving end is not important, only the number of pulses transmitted is important.

The digital computer is concerned only with discrete values. Its resolving power is determined by its word length. A computer with, for instance, a 16-bit word length can sub-divide a quantity into 2^{16} discrete elements*. This means that it can break down a length of 6.50 m into elements of 0.1 mm. If the computer is required to 'understand' analog signals, such signals (electrical ones are in the form of voltages) must first be converted into digital signals by a special device, a so-called analog–digital converter (Section 3.3.2.6). Conversely a digital–analog converter will convert digital signals into a continuous electrical signal suitable for input to analog recording instruments.

Measuring transducers (e.g. electrical resistance strain gauges) are nearly always analog appliances†. The analog representation and recording of physical quantities is simpler, cheaper and quicker than digital; the transducers are therefore much smaller and less affected by time lag. On the other hand, digital measured values have the advantage of greater reliability and freedom from interference when transmitted over long distances, but this is achieved at the expense of speed. For example, the digital transmission of a picture may take several minutes.

For special purposes, digital measuring transducers are indispensable. In this connection, special mention must be made of geometric pulse transmitters, including more particularly linear coders and angle coders.

The term 'interface' is used to denote the channels and control circuitry that connect a central processor and its peripheral units (devices which can be operated under computer control). This is illustrated in Figure 3.34 and will be further considered in a subsequent section (Section 3.3.2.6).

3.3.2 ANALOG MEASURING TRANSDUCERS

3.3.2.1 Mechanical–electrical transducers

Model analysis is concerned with the determination of mechanical quantities, including geometric and temperature-dependent quantities. If the advantageous features of electronic measuring techniques already referred to (Section 3.3.1) are to be fully utilised, the mechanical quantities under investigation must be represented by corresponding electrical quantities. There are a number

---

* A 'word' is a basic unit of data in a computer and consists of a pre-determined number of 'bits' (an abreviation of 'binary digits'); thus 'word length' denotes the size of a word measured by the number of digits it contains.

† In general, a 'transducer' is a device that converts energy from one form to another. Thus, a strain gauge is a transducer for converting mechanical strains into electrical signals.

of physical phenomena, or principles, that may be utilised to achieve direct conversion from mechanical to electrical quantities; these include (a) the interdependence of the specific resistance of a conductor and its state of strain, (b) magneto-elastic effects and (c) piezo-electric effects.

By application of these it is possible to transform strains into electrical signals. The purpose-designed devices used for doing this are usually quite simple, and these basic measurements enable other relevant mechanical quantities to be determined. Thus the magnitude of forces can be determined from the elastic strains measured; these forces can, in turn, give information about accelerations (if the mass of the moving body is known) from which angular velocities, etc. can be deduced.

Besides these direct methods, there are possibilities for indirectly producing electrical effects stimulated by mechanical action. An example is provided by the plate capacitor in which one of the electrodes is elastically displaced by the action of a force in relation to the other. The resulting change in capacitance can be measured electrically and the relationship between this change and the force that causes it can be established.

In general, mechanical action, whether it produces direct effects (changes in the molecular structure of the material) or indirect effects (geometric displacements), causes a change in the electrical constants of the transducer output circuit. A known voltage is applied to the system and the change in voltage brought about via the transducer indicates the magnitude of the mechanical action.

Either an a.c. or a d.c. voltage can be used. Through the function of the transducer, the overall resistance of the circuit will vary in accordance with the well-known formula:

$$Z = \sqrt{R^2 + \left(L\omega^2 - \frac{1}{C\omega^2}\right)}$$

from which it is evident that the voltage $U = Z \cdot I$ can be regulated by varying the ohmic resistance R, the inductance L or the capacitance C. According to this distinction, the following classification can be made: (a) resistance gauges; (b) inductance gauges; and (c) capacitance gauges.

By way of an example, the functioning of a resistance gauge will be described in greater detail.

3.3.2.2.   Example: electrical resistance strain gauge

In model analysis, as in materials testing generally, the electrical resistance strain gauge plays a very important part. With its aid, accurate strain measurements in any desired direction can be performed on a very short gauge length. With miniaturised assemblies of gauges it is possible to observe the behaviour of local stress variations occuring within a very small region. Because of their small size, these gauges can be affixed at practically any desired point on the structural model to be tested.

To use electrical resistance strain gauges efficiently it is necessary to have a clear understanding of how they work and so their properties will be dealt with here in greater detail.

All electrical resistance strain gauges function on the principle that the electrical resistance of conductors or semi-conductors is influenced by their state of stress. The observed change in resistance is due in part to the reduction in cross-sectional area resulting from lateral contraction (Poisson's ratio effect) and in part to changes in the molecular structure of the material when deformation takes place.

(i) *Metallic strain gauges:* In cases where extremely high sensitivity is not necessary (e.g. strains not exceeding 10^{-5}–10^{-6}), metallic strain gauges are preferred because of the linearity of their readings and because of their low degree of temperature dependence.

The relation between electrical resistance and strain will be explained with reference to Figure 3.35.

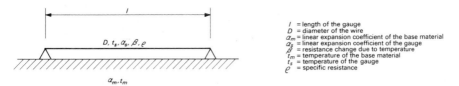

l = length of the gauge
D = diameter of the wire
α_m = linear expansion coefficient of the base material
α_g = linear expansion coefficient of the gauge
β = resistance change due to temperature
t_m = temperature of the base material
t_g = temperature of the gauge
ρ = specific resistance

Figure 3.35 Electrical resistance strain gauge.

Factor k (sensitivity)

We shall not consider the effect of temperature initially. The resistance of the wire is then:

$$R = \rho \frac{l}{F} = \rho \frac{4l}{\pi D^2}.$$

The change in resistance is found by differentiating this expression, all the quantities ρ, l and D being assumed to be variable. Furthermore, all the variables involved are assumed to be linear. We thus obtain:

$$\frac{\Delta R}{R} = \frac{\Delta \rho}{\rho} - \frac{2\Delta D}{D} + \frac{\Delta l}{l} \quad \text{and putting} \quad \frac{\Delta D}{D} = -\mu \frac{\Delta l}{l}$$

$$\frac{\Delta R}{R} = \frac{\Delta \rho}{\rho} + (1 + 2\mu)\frac{\Delta l}{l} = k\frac{\Delta l}{l} = k \cdot \varepsilon$$

whence we derive:

$$k = 1 + 2\mu + \frac{\Delta \rho / \rho}{\Delta l / l}. \tag{3.3}$$

This factor k is the sensitivity of the metallic conductor used in the strain gauge and is its most important characteristic quantity. It indicates the relationship between the change in electrical resistance and the mechanical strain at

constant temperature. Although this is not directly apparent from the above equation (3.3), it has been established that for metallic strain gauges k can with reasonable accuracy be taken as constant.

If the strain of the conductor had no effect on its specific resistance, k would be determined solely by the mechanical properties of the conductor (wire or metal foil). For $\mu = 0.3$, for example, the sensitivity would be $k = 1.6$. In practice the values of k lie in the range 2.0–2.1.

Influence of temperature

If the model and the strain gauge attached to it are heated to different temperatures, they would, if they could expand independently of each other, undergo elongations

$$\varepsilon_m = \alpha_m t_m \quad \text{and} \quad \varepsilon_s = \alpha_s t_s.$$

However, since the gauge is actually bonded to the material of the model it is subjected to a mechanical strain

$$\varepsilon = \varepsilon_m - \varepsilon_s = \alpha_m t_m - \alpha_s t_s$$

which, on taking β into account, produces the following change in electrical resistance:

$$\frac{\Delta R}{R} = \beta \cdot t_s + k(\alpha_m t_m - \alpha_s t_s).$$

The change in resistance per degree centigrade is therefore:

$$\frac{\Delta R}{R}/°C = \beta + k\left(\alpha_m \frac{t_m}{t_s} - \alpha_s\right). \tag{3.4}$$

The temperature change of the strain gauge may arise through a variety of causes which must be clearly distinguished in order to decide precisely which part of the change in resistance due to mechanical strain is of interest:

1. The model and the strain gauge, i.e. the whole test set-up, are uniformly heated or cooled as a result of normal gradual changes in the room temperature. For this case, equation (3.4) simplifies to:

$$\frac{\Delta R}{R}/°C = \beta + k(\alpha_m - \alpha_s) = k\alpha_t.$$

A rise in temperature of 1°C will produce a reading corresponding to an *apparent* strain equal to

$$\alpha_t = \frac{\beta}{k} + \alpha_m - \alpha_s$$

which has to be eliminated from the overall measurement. There are two ways in which this can be done:

(a) α_t can be eliminated by introducing a dummy gauge into the measuring circuit (Section 3.3.2.3).

(b) In circumstances where, for technical reasons, compensation of the temperature effect in the circuit is impracticable, it would be desirable to have gauges for which both $\alpha_m - \alpha_s$ and β/k were zero. By using special alloys it is indeed possible to construct strain gauges which largely fulfil this requirement. Strain gauges for special purposes are commercially available for which the value of α_1 is kept within very narrow limits ($\pm 1\mu/m\,°C$) provided that the gauge is fixed to the specified base material on which it is intended to be used.

2. Individual strain gauges on the model are briefly heated or cooled by external influences while the temperature of the base material (the model itself) does not undergo any corresponding change. Air draughts, wetting or the action of sunshine on individual gauges may produce very noticeable temperature effects on the results of the measurements. The only remedy under these circumstances is to protect the measuring positions (or the whole test set-up) from such influences which are otherwise liable to upset the results, particularly in the case of static measurements.

3. When a current flows through the conductor of the strain gauge, heat is evolved (Joule effect). Some of this heat is given off as radiation but most of it is removed by conduction into the base material. A temperature gradient thus occurs between the strain gauge and the base material and also within the base material itself. The gradient is inversely proportional to the thermal conductivity of the base material. When the current through the strain gauge is switched on, the gauge itself will at first heat up until its temperature is high enough to produce a steady flow of heat through the material. The thermal energy flow will then be in equilibrium with the electrical energy supplied to the gauge. The temperature field in the base material will give rise to a corresponding strain field which is superimposed upon the mechanical strain. It is obvious that this 'heating effect' of electrical resistance strain gauges becomes smaller in duration and magnitude as the thermal conductivity of the base material improves. In this respect, metals which are good conductors of heat behave much more favourably than, for example, plastics.

Cross-sensitivity

Wire strain gauges are composed of long, extremely thin (*ca.* 0.02 mm diameter) wires wound or arranged in a zig-zag pattern to form a grid (Fig. 3.36). Because of this arrangement, the gauge possesses (in addition to its required longitudinal sensitivity) a certain amount of sensitivity in the transverse direction since the rounded ends of the zig-zag branches form short transverse conductors. The cross-sensitivity, as this effect is called, is usually expressed as a percentage of the axial or longitudinal sensitivity and is of the order of 1%. With metal foil gauges of the photo-etched type, the cross-sensitivity can be much reduced by increasing the width of the metal at the zig-zag turning points (right-hand diagram in Fig. 3.36).

(ii) *Semi-conductor strain gauges*: With semi-conductors, it is possible to construct strain gauges of far greater sensitivity (100 times) than metallic conductors. Such gauges can be used for measuring extremely small strains or

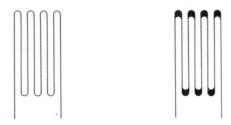

Figure 3.36 Cross-sensitivity.

alternatively larger strains without having to amplify the current to obtain suitable readings.

The main drawbacks of semi-conductor strain gauges are their high temperature sensitivity and the non-linearity of the resistance–strain relation.

The characteristics of such a gauge cannot therefore be represented by a simple factor k. The mathematical expression for the resistance change of the gauge can, however, be approximated by expansion into a series:

$$\frac{\Delta R}{R} = k_1 \frac{\Delta l}{l} + k_2 \left(\frac{\Delta l}{l}\right)^2 \cdots k_n \left(\frac{\Delta l}{n}\right)^n.$$

For practical purposes only the first two terms of the series are usually considered, and the two factors k_1 and k_2 thus serve to define the resistance–strain characteristic of the gauge. For the measurement of very small strains it is often sufficient to consider only the factor k_1, i.e. only the first term of the series.

3.3.2.3   Strain gauge circuits

The change in electrical resistance of the gauge caused by strain is always very small in comparison with the inherent resistance. For example, if a gauge with a resistance of 100 Ω is used to measure a strain of 50 microstrain (μm^{-1}) and has a sensitivity $k = 2$, the change in resistance is $10^{-4} \Omega$. To obtain useful measurements, variations of this order of magnitude must be determined with an accuracy of 1 %. To achieve this, the gauges have to be arranged in a suitable measuring circuit. This is particularly necessary because variations in the nominal resistance values of a number of gauges, all of which have the same apparent resistance, are greater than the actual resistance change to be measured.

The Wheatstone bridge circuit which is commonly used for the measurement of small changes in resistance is illustrated schematically in Figure 3.37.

A voltage U is applied to the bridge. The 'diagonal voltage' U_d (the output voltage from the bridge) is indicated by the instrument G.

By suitably choosing (or varying) the four resistances R_1, R_2, R_3 and R_4 it is always possible to make the output voltage zero, i.e. $U_d = 0$. The condition for this will now be derived.

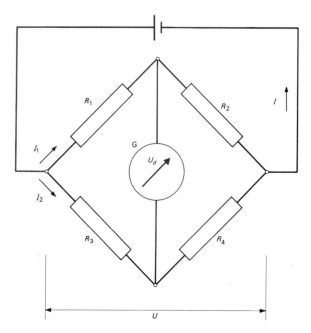

Figure 3.37 Wheatstone bridge.

For the output voltage to become zero, the voltage drop U_1 across R_1 must be equal to the voltage drop U_3 across R_3, and similarly U_2 must be equal to U_4. Hence:

$$U_1 = U_3 = R_1 I_1 = R_3 I_3$$
$$U_2 = U_4 = R_2 I_1 = R_4 I_3.$$

In addition, the total voltage drop across each of the two branches must be equal to U:

$$U_1 + U_2 = (R_1 + R_2)I_1 = U$$
$$U_3 + U_4 = (R_3 + R_4)I_3 = U$$

or:

$$(R_1 + R_2)I_1 = (R_3 + R_4)I_3 = (R_3 + R_4)I_1 \frac{R_1}{R_3}$$

whence we obtain the condition for zero output voltage:

$$\frac{R_2 R_3}{R_1 R_4} = 1. \tag{3.5}$$

If three of the resistances are of a pre-determined magnitude, it is always possible to choose the fourth so that the galvanometer G gives a zero reading, i.e. zero output voltage.

R_1 will be considered as a variable resistance, e.g. a rheostat. When R_1 is given the particular value R_0, the bridge is in balance ($U_d = 0$). If the value of R_1 deviates from R_0, the current indicated by the measuring instrument will provide a measure of R_1 (Fig. 3.38).

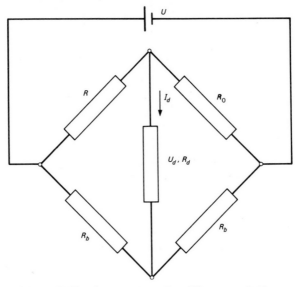

Figure 3.38   Output voltage from Wheatstone bridge.

The relationship between the reading given by the instrument and the magnitude of R_1 will be considered in a little more detail. For this purpose, let us simplify the bridge by assuming symmetrical resistances: $R_2 = R_0$; $R_3 = R_4 = R_b$. The variable resistance R_1 will, for convenience, be designated simply as 'R'.

As already stated, the Wheatstone bridge can be used for the measurement of very small resistances. For this purpose we shall introduce the relative change of resistance:

$$r = \frac{R - R_0}{R} = \frac{\Delta R}{R}.$$

We may then write:

$$R = R_0(1 + r). \tag{3.6}$$

U is the input voltage of the bridge, R_a is the resistance of the measuring instrument connected to the bridge, U_d is the output voltage through the instrument. If an amplifier with high-impedance input is connected, the output voltage is:

$$U_d = U\left(\frac{R_0(1 + r)}{R_0 + R_0(1 + r)} - \frac{R_0}{2R_b}\right)$$

or:

$$U_d = \frac{rU}{4 + 2r}.$$

r is normally very small (less than 1 %), so that we may write as an approximation:

$$U_d = \frac{rU}{4}. \tag{3.7}$$

For a strain of 1‰, the approximation involves an error of 0.5‰. If all four resistances of the bridge are assumed to be variable, a total voltage reading due to the changes in resistance r_1, r_2, r_3 and r_4 will be obtained which may be put equal to:

$$U_d = \frac{U}{4}(r_1 - r_2 + r_3 - r_4)$$

or, on replacing the relative resistances by the strains:

$$U_d = k \cdot \frac{U}{4}(\varepsilon_1 - \varepsilon_2 + \varepsilon_3 - \varepsilon_4).$$

We thus see that with the Wheatstone bridge circuit it is possible to determine sums and differences of strains. This property of the circuit can be utilised in many different ways in connection with measuring techniques.

Three commonly employed bridge arrangements are illustrated in Figure 3.39. The black rectangles represent electrical resistance strain gauges while the white rectangles are resistances of fixed magnitude which merely serve to complete the bridge circuit.

In Case 1 only one arm of the bridge contains a gauge. This arrangement is employed for dynamic and static strain measurements where temperature compensation in the circuit is not critical.

Case 2 may be utilised advantageously for many purposes. Frequently one of the two gauges is arranged to function as the *active* gauge affixed to the model while the gauge in the other arm is arranged as a *dummy*, i.e. it is fixed to a piece of the same material as the model but not subjected to mechanical strains. This bridge arrangement provides temperature compensation: if the model and the piece of material serving as the base for the dummy gauge

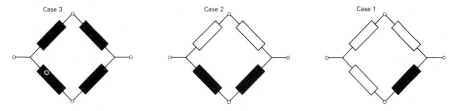

Figure 3.39   Three bridge arrangements for strain gauges (black rectangles).

undergo the same temperature changes, the temperature effect on the strain measurements will be cancelled out. For this reason, the dummy gauge is sometimes referred to as a 'compensating gauge'.

In some circumstances both gauges can serve as active gauges. For instance, if we are interested only in the flexural stresses in a shell structure subject to bending and membrane stresses, the two gauges are fixed opposite each other on both sides of the shell. Here, too, there is compensation for the temperature effect and moreover the strains associated with the membrane stresses are eliminated from the measurement. The measuring instrument now measures double the magnitude of the extreme fibre stress due to bending.

The bridge arrangement represented by Case 2 is also often employed in purpose-made ready-to-use measuring instruments. A general example is provided by the strain-measuring system incorporated in an instrument for the measurement of structural bearing reactions (Fig. 3.40). The four resistances connected in the first arm of the Wheatstone bridge are identical strain gauges, their overall resistance being R. The two gauges in each series-connected pair are mounted opposite to each other on the measuring tube of the instrument, their change in resistance being proportional to the sum of the extreme fibre strains. These gauges do not therefore measure flexural effects but only the required normal stress. The second pair of gauges acts in a similar manner in a direction perpendicular to that of the first pair. A dummy gauge for temperature compensation is mounted on a part of the instrument where it is not subject to mechanical strain (e.g. on the casing) but as close as possible to the four active gauges.

Case 3 illustrates a bridge arrangement with a gauge in each of the four arms. It provides the possibility of additional temperature compensation and is particularly useful in circumstances where major temperature influences are

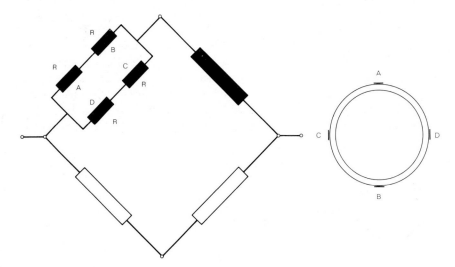

Figure 3.40   Circuit used in a measuring instrument.

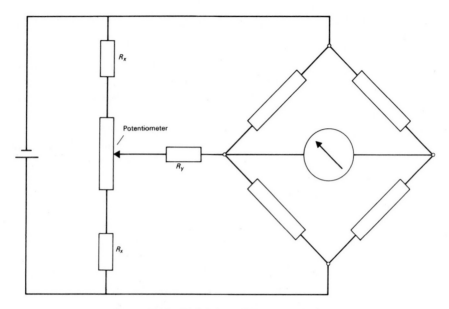

Figure 3.41   Null-balance Wheatstone bridge.

likely (e.g. in wind-tunnel measurements). It is however substantially more expensive than the arrangement illustrated in Case 2.

The relative magnitude of the changes in electrical resistance which have to be measured are of the order of 10^{-5}. In practice, the actual resistance of a strain gauge is accurate only to within $\frac{1}{100}$ to $\frac{1}{1000}$ of the nominal value so that the Wheatstone bridge is never completely in balance. A further factor which contributes to this imbalance is the difference in the lengths and characteristics of the connecting wires. As a rule therefore (but see Section 4.2.1), a zero reading on the measuring instrument will correspond to a non-existent mechanical strain value. It will accordingly be necessary to balance the bridge circuit by some special arrangement.

One of the various possible balance circuits, incorporating a potentiometer, is shown in Figure 3.41 (null-balance Wheatstone bridge).

In the above discussion the circuits described have been for resistance measurement using direct current. With alternating current the principle remains the same but capacitance and induction gauges may be used instead. In addition to the resistance it is then also necessary to balance the capacitance (by means of a variable capacitor), i.e. bring it into phase.

3.3.2.4   Systems for the measuring positions

This section is only peripherally concerned with the actual work of 'measuring'. It is concerned more especially with controlling the measuring operations. In Figure 3.34 the selector switches for the measuring positions

(more specifically, the strain gauge positions) are shown diagrammatically outside the actual set of measuring instruments.

In model testing and analysis, as indeed in other fields of scientific and technological research, it is desirable to observe as efficiently as possible a large number of measuring positions simultaneously and without loss of accuracy, and to continue such observations over long periods of time. To fulfil all these requirements would necessitate extremely expensive systems. For practical purposes, it is generally necessary to effect a compromise between the ideal solution and what can be reasonably afforded in terms of cost. Which of the above-mentioned criteria must take priority over the others will depend on each particular case in question.

The ideal technical solution is also the simplest and most reliable: each measuring position has its own circuit with current supply, balancing unit, amplifier and analog–digital converter. A computer interrogates a buffer store (a means of temporarily storing data) in which the measured data are held available and from which they are retrieved at regular intervals corresponding to the cycle time, i.e. the time it takes the computer to perform a complete cycle of operations. The accuracy of measurement is limited only by the quality of the equipment employed. Transmission of an individual item of measured data is accomplished in something of the order of a millionth of a second, so that many measurements can be stored and retrieved within a thousandth of a second. However, the ideal solution suffers from one major snag: the cost. Hundreds of amplifiers, one for each of the many measuring positions required, would cost a fortune. For most purposes it is therefore necessary to seek other ways and means.

It is obviously essential to avoid having to provide one amplifier for each measuring position. This can be done by sequential interrogation of the measuring positions. A selector device successively feeds the signal (a voltage) from each of the measuring positions to the amplifier input. To accomplish these switch-over operations takes a certain amount of time, so that true simultaneous sampling is not achieved, but this delay is acceptable in many applications. The interrogation speed can be appreciably increased by using two or more amplifiers functioning with time overlap.

The operating procedure involving a single measuring amplifier will now be considered in somewhat more detail. Apart from purely manual control, the simplest and cheapest method of 'interrogating' a number of measuring positions consists of successively selecting the circuit belonging to each by means of a step-by-step relay. The interrogation rate can be varied but is kept constant during the measuring operation itself. In each of its successive working positions the step-by-step relay energises the selected data channel relay and thereby causes the measuring signal (output voltage from the measuring device) to be fed to the amplifier. With this technique the measured data can be recorded at fixed time intervals in a given cyclic sequence (and only in one direction). Usually the relays employed for establishing the electrical connections with the measuring positions are of the ordinary mechanical type, with a switching time of about 20 milliseconds. In this way it is possible to attain interrogation rates of up to *ca.* 5–10 per second.

If a computer is available for monitoring the experimental procedure, the interrogation of measured data can be performed much more elegantly. At the present time a variety of so-called scanners are available which, in conjunction with the computer, enable the measuring positions to be selected with complete freedom of choice, both with regard to time and to location. With this freedom a notable increase in the flexibility of testing can be achieved in addition to clearing the way for measuring techniques of greater precision (Section 3.3.3). Thus the measuring operations are not only controlled in accordance with programs that are adaptable to the requirements of any particular test but can also be regulated and modified with reference to the results obtained while measuring is actually in progress.

The relays used for closing the measuring circuits selected by the computerised equipment are of the type called 'reed relays' with switching times of the order of 1 millisecond. A typical reed relay consists of two flexible magnetic tongues (with contacts at their ends) sealed into a glass tube with protective gases or a vacuum; the tube is mounted inside a solenoid which is energised to operate the contacts. The relays are operated by the scanner via transistors. The functions and sequences involved in the reception of a single measured value by the computer will now be described with reference to Figure 3.42. The signal representing the measured value is passed to a digital voltmeter (Section 3.3.2.6) and then fed to the computer.

(i) From the program for the model test, the computer seeks out the instructions applicable to the specific measuring position (its channel, measuring range of the voltmeter, type of measurement and scan time). It takes the computer less than 1 millisecond to pass these instructions to the equipment concerned and it is then free to perform other duties (Section 4.2).

(ii) The scanner connects the selected data channel to the voltmeter.

(iii) A programed time interval occurs to allow for slight inertia delay in the movement of the reed relay tongues or to wait for a specific instant of time which is technically desirable in connection with the test.

(iv) A time interval is made available for automatic resetting of the measuring range in the event of this range being exceeded by the incoming measuring signal.

(v) A time interval of 100–500 microseconds is introduced if alternating current signals are measured with the aid of a voltage–resistance converter (in combination with inductance or capacitance strain gauges).

(vi) A so-called gate time for the voltmeter occurs. During this interval (e.g. 10, 100 or 1000 microseconds) the voltmeter estimates the input voltage. The resolution of the measured value and the elimination of hum depend on this time interval.

(vii) When the voltmeter has signalled its readiness to transmit the data, a waiting time of 1 microsecond is allowed for the transmission lines to clear. The information is then fed into the working store of the computer.

(viii) Conversion of the measured data into floating point representation is undertaken by the computer. (This is a method of representation in which a number is represented by two sets of digits, known as the fixed point part and the exponent or characteristic.)

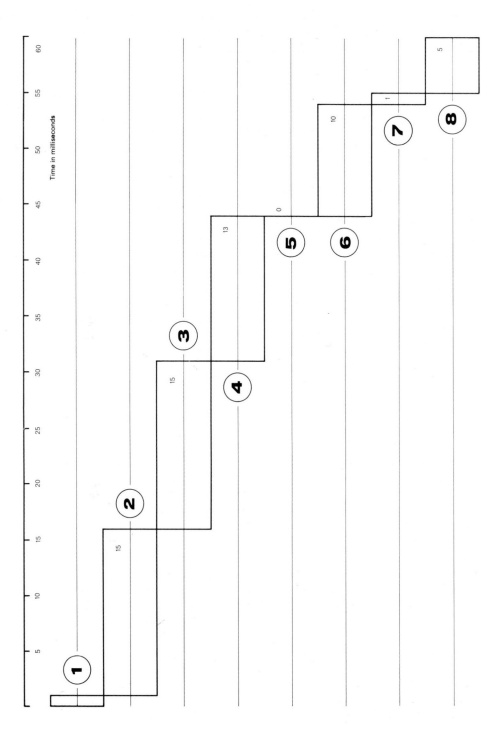

Figure 3.42 The steps involved in the reception of a measured value by the computer.

This sequence of stages in the transmission of data (which, as envisaged here, takes at least 60 milliseconds to accomplish) must be regarded as an example associated with a particular set of equipment. It does, however, illustrate the various general problems that are encountered.

3.3.2.5 Amplifier and power supply to the strain gauge bridge

In strain measurement the measuring bridges are operated at between 0.5 and 10 V. If it is desired to measure strains down to a magnitude of 10^{-6}, then, assuming a bridge input voltage of 1 V and a sensitivity factor $k = 2$, the output voltage will be $\Delta U = \frac{1}{4} \times 2 \times 10^{-6} \times 1 = 0.5 \times 10^{-6}\,\mathrm{V} = 0.5\,\mu V$. This signal is too weak to give a galvanometer reading and too inadequate to operate a recording instrument. It must be amplified, and for this purpose many different makes of amplifier are commercially available. In principle, these devices are precision voltmeters for very low voltages.

The instruments developed for performing the duties envisaged here generally comprise, in addition to the actual amplifier, the following elements: (a) a power supply unit for the measuring bridge with a voltage selector; (b) fixed resistors for connecting into the bridge arms in combination with resistance strain gauges, to provide various circuit arrangements (cf. Fig. 3.39); (c) a balancing unit for one bridge normally provided as an integral feature of the apparatus; (d) output for a galvanometer with visual indication and a further amplified output for connecting to analog recording devices; and (e) readings given directly in strain units; the factor k can be varied on many devices of this kind.

A distinction must be made between four basic types of measuring bridge; alternating current or direct current; direct-readout bridge or null-balance bridge (compensation principle).

The advantages and disadvantages, from the electronic engineering point of view, of using alternating current or direct current have been extensively debated. For the present purpose, however, the main features of these two techniques will be compared only as far as they affect the applications:

Alternating current	*Direct current*
Inductance gauges and capacitance gauges can be connected directly into the circuits.	Only resistance measurements are directly possible; inductance gauges and capacitance gauges require suitable transducers (more expensive).
The capacitance of the connecting leads principally affects the measured value.	Only the resistance of the leads affects the measured value; can be easily controlled.

Alternating current	*Direct current*
Influence from external inductive disturbances (hum). The higher the carrier frequency, the more sensitive are the measurements to disturbing influences.	No inductive and capacitive disturbances are possible.
Temperature effects play no part.	Temperature effects may be very noticeable. Great caution is necessary.
Measuring rate is limited.	Suitable for very rapid data collection.

The functioning principle of the direct-readout (or direct-indicating) Wheatstone bridge is apparent from the explanations and derivations given in Section 3.3.2.3. The non-linearity of the system involves some error in the measurements but normally this error is not large enough to be significant. This technique has the advantage of giving immediate and direct indication of changes in the measuring circuit.

The functioning of the so-called null-balance bridge (for measurements on the compensation principle as opposed to the direct-indication principle) can be understood by reference to Figure 3.41. The galvanometer connected to the amplifier serves here merely as a 'null indicator'. Instead of taking a reading of the output voltage (or the strain), the operator varies the setting of the potentiometer and thereby balances the bridge, i.e. restores the output voltage to zero (zero deflection on the galvanometer). The angle of rotation of the control knob of the precision potentiometer is adopted as the measure for the quantity to be determined. Theoretically this reading is always strictly proportional to the measured voltage. An accurate null-indicator instrument with amplifier is much easier to design and construct than an instrument that can give strict linearity in its readings. The equipment is therefore accurate and relatively cheap and can be used advantageously in all circumstances where a high speed of measurement is not essential, i.e. generally speaking, in static as distinct from dynamic strain gauge applications.

The automation of rapid measuring sequences, calls for direct-readout bridges. In Section 3.3.3 a method is described whereby the drawbacks of this technique are overcome.

3.3.2.6 Analog–digital converter

The analog–digital converter (for the present purpose more specifically a digital voltmeter) occupies a key position in the chain of equipment extending from the electrical transducers (usually strain gauges) to the computer. The development of these converter units has facilitated the processing of the analog transducer output signals by means of the digital computer.

As the name implies, a converter of this type converts analog signals into digital signals. Initially, efforts were directed only at producing a digital indicating instrument for voltages. The conventional method of indicating a measured value by means of a needle or pointer giving a deflection on a scale was to be replaced by digital indication which would permit speedier visual recording.

The process of conversion consists in comparing the analog input voltage for the computer with a number of exact reference voltages which are generated within the computer and whose values correspond to digital quantities. With a binary-coded decimal notation*, for example, the maximum voltage can be resolved into 999 parts with a total of $3 \times 4 = 12$ reference voltages with values of 1, 2, 4, 8, 10, 20, 40, 80, 100, 200, 400, 800. Each group of four reference units corresponds to a decimal place, and the 10 requisite combinations of the four bits (= binary digits) correspond to a decimal digit. The representation of a measured quantity (369) by the 12 reference voltages is shown in Figure 3.43. The machine first automatically switches to the reference voltage of 100 and compares this with the incoming signal. If the reference voltage is smaller than the signal, the machine tries the next higher voltage, i.e. 200, and so on. If the sum of the reference voltages switched on exceeds the measured value, the machine then reverts to the preceding lower value (in this example, 300) and now progressively increases the reference voltages in smaller increments (10, 20, etc.) until the measured value is again exceeded. Finally, the procedure is repeated with the smallest increments (1, 2, etc.). Each time the result is stored in the memory of the computer, and on completion of the selection process described here a readout of the final result is given.

If a binary, as distinct from a decimal, readout will serve the purpose, the BCD coding can be dispensed with. In that case purely binary representation with the same degree of resolution can be achieved with 10 reference units, each corresponding to a binary digit (e.g. a display on the front panel of the machine). This makes for greater simplicity and, in general, binary digital data processing equipment is cheaper and faster than comparable decimal equipment.

The decimal readout is advantageous where substantial numbers of measured values have to be read visually. However if measurements must be

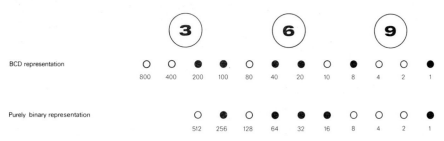

Figure 3.43   BCD representation.

* BCD code: a method of using groups of binary digits to represent decimal numbers, with each digit position of a decimal number being allocated four bits.

performed in rapid succession this becomes physically impracticable. In such circumstances the digital values are fed direct through suitable electronic connecting units to a recording instrument or to the computer for further processing. In the latter case the binary representation is advantageous.

Digital voltmeters as described above operate on the principle of successive approximation. There is, however, a type of instrument for converting analog to digital signals which functions on a completely different principle: so-called digital integrating voltmeters. In an instrument of this type, a high-frequency electrical oscillation is generated, its frequency being proportional to the incoming analog voltage (the measuring signal). Assuming that the signal varies with time, the following relation is valid:

$$\frac{1}{T} = \int_0^T v(t)\, \mathrm{d}t = \frac{C}{T} \int_0^T U(t)\, \mathrm{d}t = \frac{C}{T} \bar{U}$$

or

$$\bar{U} = \frac{1}{C} \int_0^T v(t)\, \mathrm{d}t.$$

In this expression the right-hand integral is (accurately to one cycle) equal to the sum of the cycles of the oscillation during the time T. The measurement is thus reduced to a numerical counting process performed by an electronic counting device. At the end of the predetermined length of time the sum of the cycles counted by this device corresponds to (i.e. is proportional to) the mean value of the voltage.

This method has the disadvantage that the speed of performing the measurement is limited by the requisite measuring time T, but against this it has the major advantage of determining mean values.

Every measuring signal (including the signals obtained from static quantities) is affected by hum, i.e. periodic disturbances are to a greater or less extent superimposed upon its nominal value. The disturbed signal may be 'cleaned-up' by determining the mean value. The measuring time adopted with integrating voltmeters is usually a multiple of the cycle of the mains alternating current,

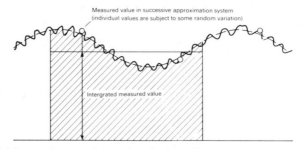

Figure 3.44 Comparison between two operating principles for digital voltmeters.

as disturbances in the measuring circuit are liable to arise from inductive influences due to the magnetic fields set up by power installations (see Figure 3.44).

3.3.2.7 Interface

The term 'interface' in computer technology refers, in general, to a common boundary between two systems, especially data processing systems. More particularly, it may denote the channels and associated circuitry that form the connections between the computer and its peripheral units.

Every type of computer operates with standardised electrical and mechanical units which are adapted to the electronic components of which the computer is constructed. Thus it makes use of strictly demarcated voltage ranges for distinguishing between 'yes' and 'no' or between 'plus' and 'minus', e.g. from $+6$ to $+12$ V for 'yes' and from -1 to -5 V for 'no'. Its working rhythm is determined by an oscillator of constant frequency. All operations are performed in whole multiples of the period of that oscillation (e.g. 0.2 microsecond). Thus the core storage cycle time—the time the computer takes to retrieve an item of data from the core storage and being it into relationship with some other item of data held in the register by means of a machine operation—may be eight periods in length (i.e., 1.6 microseconds in our example). The working storage is able to take in data only during specific, periodically repeated lengths of time. The construction of the computer is to a great extent determined by its word length and by the code with which it internally processes numerical values. Operations between the arithmetic unit, registers and core storage are always performed with the full number of bits (binary digits) determined by the word length. If only some of the bits have a meaning allocated to them, all the others will have to be made zero.

Since the peripheral units seldom present their data in a form that is completely compatible with the mode of functioning of the computer, the interface must provide the necessary facilities to enable communication between those units and the computer to take place. More particularly, the incoming signals must be converted and adjusted to the standards of the computer and vice versa.

Thus the voltage levels have to be adapted, often involving a change of sign (positive or negative logic).

The phenomena or occurrences detected by, or transmitted to, the peripheral units are seldom periodic in character and certainly do not coincide with the operating period of the computer. Accordingly, the interface equipment must comprise some kind of memory or storage device (flip-flop register) in which the asynchronously incoming signals from the computer and the peripherals are stored until the receiving unit is ready to take the information. Depending on the type of computer concerned, this information holding time in the intermediate storage device may merely be a few microseconds.

Figure 3.45 Control unit (with punched-tape input) for the loading apparatus described in the text (see also Figure 3.26). The unit is connected to a fixed-program digital computer supplied by the firm of Heinrich Dietz. This computer also monitors data gathering and provides output of the measured results on a Kienzle printer and an IBM 016 card punch.

In addition to data, the interface also transmits addresses (indications for identifying a unit, such as a register, where specific information is stored) and instructions.

Sometimes the interface equipment operates on the interrupt principle. It may happen that during the execution of a job the computer has to communicate with several peripheral units and may additionally be engaged in performing arithmetical operations. These various activities must therefore each be assigned a priority. Units which yield data in rapid succession (e.g. magnetic tape or measured data from an analog–digital converter) must be given higher priority than those which give a slower input of data (e.g. teleprinters or the control data of a mechanical process). Thus when a rapid input unit

signals that it has measured data ready for transmission to the computer, the latter will interrupt the lower-priority program on which it is then engaged and take in the data waiting in readiness in the interface register.

3.3.3   COMPARISON METHOD

3.3.3.1   Principle

In the above discussion of measuring techniques, we encountered the principle of compensation in the use of dummy gauges to eliminate temperature effects (Section 3.3.2.3). In a more general form, the comparison principle is utilised in the null-balance bridge for measuring with greater reliability and precision (Section 3.3.2.5). A feature common to both examples is that the electrical quantity to be determined is compared with a known reference quantity. In the first of these two cases, this is done merely to cancel out disturbing influences; in the second, the reference quantity serves as the standard for determining the actual quantity to be measured. It is possible to take this a stage further and to compare directly not only the electrical but also the mechanical quantities that we wish to determine. The extension of the compensation principle is designated the *comparison* method. In suitable cases, it is possible by this method to attain far superior accuracy of measurement independent of temperature and other incidental conditions, while imposing less exacting requirements as to the quality of the electrical measuring instruments employed.

3.3.3.2   Manual comparator

The determination of the stresses in a structural model made of plastic by means of electrical resistance strain gauges is complicated by the fact that the modulus of elasticity of plastic is a time- and a temperature-dependent quantity. This means that the strain measurements, from which the magnitude of the stress has to be deduced with reference to that modulus, are liable to be inconsistent.

In the procedure to be described below, however, the stress is not determined from strain measurements with their inherent uncertainty. In this case, the apparatus 'weights' the stress directly (Figures 3.46 and 3.47). A reference bar made of the same material as that of the model is supported under statically determinate conditions and is subjected to a bending moment of known magnitude at one of its bearings. The moment is produced by a weight which can be moved along a balance beam and its magnitude can be read from a scale. Thus the stress in the extreme fibres of the reference bar is known for any position of the sliding weight. An electrical resistance strain gauge is now fixed to the reference bar (at a particular location on the bar) and is connected

Figure 3.46   Manual comparator (Author's system).

Figure 3.47   Rear view of the comparator with cover removed to show the sliding weight for producing the bending moment and the reference bar with statically determinate support and dummy gauge.

as the dummy gauge in a bridge arrangement, as in Case 2 of Figure 3.39, while the active gauge is fixed to the model. Before measurements commence, the bridge is balanced while there is no load on the model and while the manual comparator is at zero. Now if a load were only applied to the model, the galvanometer reading would correspond to the strain affected by the creep of the plastic, i.e. it would vary with time. With this technique, however, the operator of the equipment can, immediately after a load is applied to the model, shift the sliding weight on the balance beam of the comparator until the galvanometer reading is returned to zero.

If the model and the reference bar are thus loaded simultaneously, the strains which occur at the measuring positions on both model and bar will traverse identical creep curves. The exact value of the stress can now be read directly from the scale on the comparator. Since the modulus of elasticity has thus been eliminated from the measuring procedure the results obtained are also completely independent of the ambient temperature. Of course, it is not possible to apply the loading to the model and to the reference bar absolutely simultaneously. Yet experience with the apparatus has shown that in practice complete stability of the reading is obtained within a few seconds. This has proved to be the most accurate and reliable method of stress measurement for use with electrical resistance strain gauges. However, the manual comparator is not a very suitable element in a fully automatic data collection system.

It is too slow for the purpose, even if it is automated by means of a servomotor system.

3.3.3.3   Generalisation of the method

In model testing sometimes hundreds of thousands of measured values have to be collected. The analog signals, which follow one another in rapid succession at extremely short intervals, are converted to digital signals and fed to the computer. The high speed with which the signals can be gathered from the gauges, the freedom in programing the sequence in which this is done (Section 3.3.2.4) and the utilisation of the computer as a temporary data store are factors which make it possible to perform a comparison of the measured value with the mechanically produced reference value practically simultaneously within the computer itself. In an automated measuring system the correction effects (similar to those obtained with the manual comparator) are achieved by taking a strain measurement on the reference bar (ε_0) immediately after each strain measurement on the model (ε_M) has been stored and instantly comparing the two. The difference in time between the two measurements is insignificantly small. Let σ_M be the stress to be determined in the model and let σ_0 be the known reference stress in the bar. The computer each time thus determines

$$\sigma_M = \sigma_0 \frac{\varepsilon_M}{\varepsilon_O}.$$

It must of course be borne in mind that σ_O is a *comparison stress* which will be identical with the actual stress at the measuring position on the model only if a uni-axial state of stress exists there. If not, the measured stress will have to be corrected by the methods of rosette analysis to take account of Poisson's ratio. The main objective will however have been achieved: the modulus of elasticity has been eliminated from the evaluation of the measurements. There are other advantages, too. Since both ε_M and ε_O are measured with the same set of electrical devices (amplifier, analog–digital converter) and only the quotient of the two strain measurements is used for arriving at the stress, it is not necessary to know the absolute value of the voltage produced by the strain (due to the change in resistance of the gauge). In other words, the result of the measurements is theoretically independent of the amplification factor of the equipment. The amplification factor is only important in determining the degree of resolution attained. This distinctive feature of the comparison method is of great practical value as it enables calibration of the electrical instruments to be dispensed with. In addition, any change in the characteristics of an instrument during measuring operations over long periods of time has no effect on the result, so that long automated measuring sequences in particular are largely immune from disturbing influences. The only requirement that the characteristic of the amplifier must fulfil is there should be proportionality between input and output voltage. As a source of error, zero drift can also be

eliminated by means of the technical resources available through 'hybrid analysis' (Section 4.2.1).

The comparison principle, based on direct mechanical comparison, need not be confined to strain measurements. It can also be applied to the measurement of other physical quantities if the necessary reference devices are available. Figure 3.48 shows a 'universal' reference device forming part of the

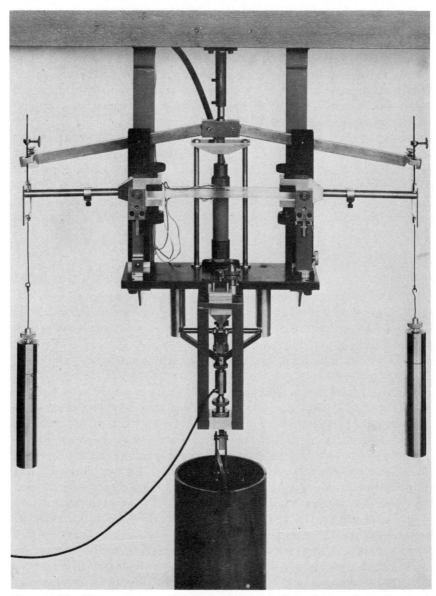

Figure 3.48 'Universal' automatic pneumatically-controlled reference device for simultaneous calibration of strain, force and displacement (Author's laboratory).

general hybrid equipment. At the top is the reference bar of Plexiglass which is loaded with a constant moment at each end by the two weights hanging down on the left and right. A larger weight (at the bottom) applies a force of known magnitude to the reference element for the measurement of bearing reactions. A standard displacement can also be produced for the calibration of inductive deflection gauges.

4.

COMPUTERS AND MODEL ANALYSIS

4.1 DATA-PROCESSING IN STRUCTURAL ANALYSIS

Even today there are a good many engineers who believe that model testing as a branch of structural engineering science is 'on the way out' if not actually obsolete as a result of the remarkable development of the electronic computer and its application to structural analysis. Misconceived though this view may be, it deserves to be treated seriously since it strikes at the essential meaning and purpose of structural model testing.

It is true that the performance and potentialities of modern numerical methods seem overwhelming. Indeed, some structural research laboratories, hypnotised by the apparently unstoppable progress of computer techniques for dealing with problems of structural analysis, have neglected their model-testing activities and in certain cases have even completely abandoned them. Yet, in this context, it is significant that in the United States, home of much of the development of the modern computer, a revival of interest in model analysis and testing is taking place.

In order to obtain a conclusive answer to the question raised here, i.e. the justification for model testing, three aspects must be considered: (a) the essential difference between the methods; (b) the relative importance of computerised analysis of structural systems; and (c) the relative importance of model analysis.

The general assertion that model analysis is doomed can easily be refuted. The answer has in fact already been given in Section 1.1 of this book. Beyond the boundary of what can be described analytically, there will always be a field of application for model analysis and testing. The essential difference between the two methods ensures that there is ample scope for both to go on developing side by side.

But what is the situation near the boundary? Undoubtedly it is this border-line region, where both the computer and the structural model strive to solve the same problems, that the sceptics chiefly have in mind when they criticise model analysis. To make a proper assessment of the situation in this important range of problems (for instance relating to the elastic behaviour of irregular shapes of slab-type structures such as bridge decks), where exact methods of solution are still lacking, is certainly a matter of some significance. It calls not only for qualitative but also for quantitative knowledge of the possibilities afforded by modern methods of structural analysis as practised on the big

computers of today, and also on those of the next generation. Before trying to make an assessment of the position, it should be pointed out that the critics invariably base their attitudes on an obsolete concept of model-research procedures. The true picture of present-day experimental structural analysis, with which the engineer at large is still all too unfamiliar, will be outlined below in Section 4.2.

Let us consider the digital computer and its potentialities as a tool for assisting the structural designer. In every sector concerned with the design of engineering structures efforts are being made to use the computer as a means of speeding up calculations and also to extend the fields of application and to devise new numerical methods. Some important applications of the computer to the analysis of structures are:

(a) Elastic theory: plane frameworks, grids, shell and plate structures for first- and second-order theory, three-dimensional elastic structures
(b) Plastic theory: the same categories of structures indicated above
(c) Ultimate load theory: reinforced concrete structures

The electronic computer has proved particularly suitable for the analysis of elastic framed structures. The analysis of a statically indeterminate system of this kind can in principle be reduced to the mathematical problem of solving sets of linear equations. In cases where very large numbers of unknowns are involved—possibly running into thousands—the task of solving such equations, while virtually impossible by manual methods, can be performed quite easily in a matter of minutes by a big modern computer. In fact, almost any structural design problem involving the analysis of elastic framed systems can be tackled by the computer. That computerised data processing has not become commonplace for dealing with such problems is simply because these machines are not yet sufficiently accessible to all who may have occasion to use them. But this state of affairs is bound to change in the near future. Framed structures have never been a field that was of much interest from the viewpoint of model analysis. The situation is different where elastic plate-, slab- and shell-type structures (which can, broadly, be referred to as 'stressed-skin structures') are concerned. Until fairly recently, model analysis often provided the only means of accurately determining the behaviour of irregularly shaped structures of this general category, i.e. those designed to any desired shape to suit the requirements of a particular situation. What has the computer to offer in connection with the design and analysis of such structures?

The 'exact' mathematical formulation (within the limits of whatever hypotheses are adopted) of problems relating to plate and shell structures leads to a set of differential equations in which the equilibrium conditions are described in terms of stresses, and the deformation conditions are described with reference to the infinitesimal element. It is only in very simple cases that such sets of differential equations can be directly integrated. In the conventional numerical methods of solution (expansion into series, method of finite differences), which can be applied quite generally theoretically, physically

exact relationships are utilised. The results converge to the true solution. For reasons which will not be detailed here, attempts to establish general programs for the analysis of plates and shells, based mathematically on these well-known numerical procedures, have met with only limited success.

With the development of the finite element method, however, this situation has changed in a fundamental manner. From a mathematical point of view, this method abandons the aim of finding an iteration procedure for solving the differential equations and instead uses a more physical concept. The plate or shell is divided into a number of finite-sized elements by dividing lines which intersect at nodes. The geometric shape of the elements adopted in a system is subject to restrictions; thus, for example, only plane triangles and rectangles may be permitted for describing even structures such as three-dimensionally curved shells. The variation in thickness within an individual element is neglected. Thus the geometry of the structure is reproduced by an approximate model which more accurately represents the true shape if a larger number of elements is used to describe it.

For each of this limited number of element types, we now deduce force–deformation relationships based on simplified assumptions regarding the stress distribution or strain distribution within the element. Depending on whether the description of the state of distortional deformation inside the elements is based on stress or on strain relationships (expressed, for example, as polynomial functions), the elements are referred to as constituting an equilibrium model or a deformation model. At the dividing line between adjacent elements in an equilibrium model there must always be continuity of stresses, and in a deformation model there must always be continuity of strains. Because of the hypothetical and simplified assumptions concerning stress and strain distribution within the elements, it is not possible in general to satisfy both these conditions.

Assuming that the character of the stress or strain functions is known, we can now unambiguously describe the elastic state in the interior and along the edges of the element entirely in terms of deformation or of equivalent forces (which are in equilibrium with the edge stresses) acting at the nodes. This is where the distinctive character of the finite element method lies: it enables the problem of analysing irregularly-shaped plate and shell structures to be reduced to solving a set of linear equations. Thus, from the computer's point of view, the problem becomes (qualitatively, at least) essentially similar to that presented by the analysis of framed structures. The main object of any further development of the finite element technique must be to extend the system of elements susceptible to convenient analysis, in order to obtain an accurate approximation of irregularly-shaped structures with elements of the largest acceptable size, i.e. with the coarsest possible network of dividing lines.

The early successes achieved by the finite element method are indeed impressive. In Figure 4.1, for example, the results of a computer analysis of a hexagonal shell with a curved middle surface and variable thickness are compared with those obtained from a model test. The analysis was performed with the aid of the STRIP program system developed by a Swedish/Swiss team.

Figure 4.1   Comparison of model test results with the results of a computer analysis with the finite element method (STRIP program system). The sub-division of the structure into finite elements used in the analytical method is shown drawn on the model.

The network employed, which consisted of triangular elements, can be seen in the photograph. The individual elements are plane and of constant thickness, but the thickness varies from element to element. For symmetric loading of the shell, by making use of geometric symmetry it is possible to confine the analysis to one-twelfth of the hexagon and to feed the data from this into the computer. The agreement between the results of the model test and the computer analysis is impressive. In fact very small differences between the two sets of results make it difficult to decide which are closer to the (unknown) theoretically exact solution.

Successful applications like the one illustrated by this example have led many structural engineers to believe that through the use of the computer it will soon be possible to solve even the most complex problems in the domain of elastic theory. Thus a theory, tailored to the computer and capable of considerable further extension, can in principle be utilised to formulate all problems encountered in this domain. Hence in purely academic terms the problem as a whole has been disposed of. However, in order to fulfil practical requirements something more than the knowledge that any particular problem can be solved is required; instead, we need a procedure which will indeed achieve a solution in specific cases. Hence in order to arrive at a practical

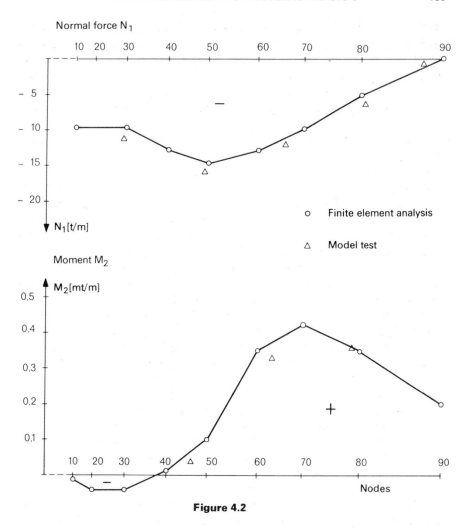

Figure 4.2

assessment of the finite element method in relation to elastic-model analysis, some knowledge of the weaknesses and limitations of the method is necessary.

It has been mentioned above that through the introduction of the concept of finite elements the analytical problem has *in principle* been reduced to the same type of problem as that tackled by the computer in dealing with the structural analysis of frames. However, when a quantitative comparison is made between the analytical problems involved in plate and shell structures and in framed structures, respectively, a different picture emerges. Broadly speaking, the number of dividing lines and nodes needed to adequately describe the geometry and structural features of plates and shells increases in proportion to the square of the number of such elements in comparison with framed systems composed of bar-like members. For instance, if it is necessary to consider, say, 20 points on a beam in order to attain a certain desired degree of accuracy, an equally accurate description of a comparable plate or shell

will require 400 points (and therefore a correspondingly large number of nodes). Geometry is not the only reason that compels us to use a vastly greater number of points to describe plate and shell structures. The analysis of framed structures is an 'exact' procedure whereas the finite element method is an approximation, and in order to attain a high degree of accuracy for the results it is necessary to employ as fine a mesh of dividing lines as reasonably possible.

The capacity of even the largest computer has its limits, and experience has shown that in connection with the analysis of plates and shells this limit is reached surprisingly soon. An important reason for the rapid exhaustion of the computer's capacity, besides the sheer number of equations, lies in the peculiarity of the numerical procedures for dealing with this type of problem. The precision with which the computer solves sets of equations depends on the word length with which it operates. During the performance of matrix operations there is a continual loss of digits as a consequence of the limitations of precision, and with any given word length a point will be reached where these 'numerical instabilities' rule out the processing of even larger matrices. It then becomes necessary to operate the computer with multiple word lengths. This reduces its capacity in an inverse ratio and moreover increases the computing time by factors of the order of 100. Through the extensive use of large peripheral stores it is possible to further increase the number of equations that can be solved. However, the computing time and therefore the cost of the operation then increase tremendously, so that one then can hardly escape the feeling that a sledge-hammer is being used to crack a nut.

Because of the differences in computers, their word lengths, construction of the central unit, nature of the peripheral equipment and type of software used, it is not possible to give a general rule for the amount of computation work required for solving a given number of equations. The computing effort increases roughly in proportion to the third or fourth power of the number of unknowns and thus quickly exceeds all reasonably acceptable limits. This characteristic has already been pointed out, in principle, in Section 1.1. From the point of view of cost, the situation can be approximately summed up as follows: if it costs £200 to sove 2000 equations, it may cost as much as £1000 to solve 4000 equations and the cost of solving 8000 equations would be prohibitive. Developments in computer technology may improve this situation, but cannot radically change it in the foreseeable future. Yet many problems commonly encountered in structural engineering practice are liable to require hundreds of thousands of elements to describe them with sufficient accuracy, for example, a multiple box-girder bridge, continuous over several spans and curved in plan.

The type of problem that the finite element method in its present state of development can neatly solve are those relating to plates and slabs, and the simpler shell problems, particularly where geometric symmetry or symmetry of loading can be utilised to simplify them.

On the other hand, model analysis can tackle all manner of problems concerned with the elastic behaviour of structures. By availing itself of the

modern experimental methods described in the following sections, model analysis can do this virtually regardless of the difficulty or complexity of the problem. If computer analysis and present-day model analysis are compared further, even more significant features emerge. The following are relevant to plate and shell structures:

(a) When the approximations on which the finite element method is based are taken into account, the results obtained from model testing are generally more accurate (Section 4.2). The result of the model test is verifiable, which cannot always be said of the analytical method.

(b) Important intermediate results of model testing (influence functions) can be stored in a form which makes it subsequently possible to obtain additional information at any time with the minimum of extra effort (Section 4.2).

(c) The effort involved in feeding data into the computer for the analytical description of the geometry and loading conditions may exceed reasonable limits in the case of complex structures. This tedious task is avoided in model testing and is instead replaced by the agreeable task of skilled technicians specialising in this type of work constructing the model.

(d) The total time requirement for model testing, starting from the preparation of the working drawing for the model up to the time when the test results are available, can be significantly shorter than the total time needed for carrying out the analysis on the computer.

A closer examination of the facts thus leads to the conclusion that, even for elastic structures, model analysis and testing still have a wide field of application with much scope for future development, which cannot be validly taken over by the computer in the foreseeable future.

4.2 HYBRID ANALYSIS

4.2.1 TECHNICAL FEATURES

The spectacular development of electronic digital computers is something of which every engineer is aware. In a period of little more than 10 years the operating speed of computers has increased a hundredfold. Thanks to miniaturisation of components, present-day mini-computers now have a higher performance than the big computers of 10 years ago. The engineer familiar with conventional analytical procedures is of course rightly impressed by such developments and, not surprisingly, tends to believe that the computer will sooner or later relieve him of all analytical calculations.

However, the computer is but one product of modern electronics, though admittedly one of the most impressive ones. There are many other applications of electronics which, taken as a whole, are at least as important to technological evolution.

The kind of equipment that has become a commonplace feature of modern life—colour television, stereophonic music, tape recording, miniaturised transistor radios, etc.—affords examples of this. The tremendous advances in the entire field of telecommunications have made a powerful impact on the pattern of social life and will undoubtedly have an even greater impact in the future. Thanks to the possibility of converting analog data into digital data (and vice versa) with lightning speed, the computer can, as it were, converse with analog measuring instruments. Control decisions based on the results of complex data-processing can be made in fractions of a second. The layman is still too little aware of the far-reaching implications of this capacity of the computer. The conventional concept of mechanical precision is undergoing a radical change. By means of skilful logic control and electronic monitoring, it is often possible with appreciably simpler mechanical means to construct cheaper, more accurate and more versatile machines.

The Apollo program of manned flights to the Moon is an outstanding showpiece for the power and performance not only of sophisticated computer equipment but also of measuring techniques and telecommunication. The components used in modern electronics function rapidly and accurately with minimal power consumption. They are versatile and small and light beyond the dreams of a few years ago. The accuracy and speed of electronic equipment

is utilised also in analog computers (Chapter 2) which, in general, are more suitable for solving linear and non-linear differential equations whereas digital computers are better equipped to deal with algebraic equations. A hybrid computer is a combination of analog and digital computing devices between which there is a constant interchange of data. Such an arrangement combines the properties of the analog system with the accuracy, memory capacity and programing flexibility of the digital system.

Model analysis presents virtually an ideal field for the application of the resources of modern electronics. It is half experiment and half calculation and requires a very high standard of measuring technique, flexible process control and intelligent data-processing. Experimental procedures in which a digital computer operates in combination with the testing and measuring set-up as a single integrated whole will be referred to as 'hybrid analysis'. In the past, model analysis has too often been looked upon as a last-ditch expedient, whose disadvantages have to be accepted in situations when there is no other way of solving the problem. Early model-testing techniques were often laborious so that in many cases their application had to be ruled out for reasons of time and cost.

Before examining the more attractive present-day possibilities, it will be instructive to review the successive stages required in carrying out a conventional elastic-model-testing project:

(a) First a model is constructed, often with relatively primitive means, and measuring transducers (strain gauges, etc.) are fixed to it. In view of the time-consuming evaluation of the results of the measurements, there is an understandable reluctance to use more than the minimum number of measuring positions.

(b) The model is usually built up on a testing frame improvised or adapted for the particular set of tests concerned.

(c) The loading devices usually leave much to be desired. A large number of weights, whose distribution on the model must be calculated and which must each be of the correct value, has to be applied. This laborious and slow operation restricts the investigation to only a few loading cases.

(d) Interrogation, or scanning, of the measuring positions can be automatic but it remains necessary to supervise and monitor the measuring operation.

(e) Evaluating and recording the results are done manually and cost additional time.

If model analysis is to shake off its reputation of laboriousness and frequently also of inaccuracy, it must shed its character of 'messing about' with gadgets and contrivances. Hybrid analysis, in which the computer is continuously engaged in dialogue with the model being tested, raises model analysis to the status of an incomparably more efficient tool of investigation and also opens up new fields of application and new possibilities for dealing with structural design problems (Section 4.2.3). The efficiency and power of this technique can be summed up as follows:

The accuracy of the measurements is greatly enhanced in comparison with manual methods. The measuring conditions (e.g. bridge voltage, temperature,

etc.) can be continuously monitored, and calibration measurements can be called for by the computer as often as may be desired. With this information the computer can continuously 'clean-up' the measured values obtained.

The quantity of measured data that can be handled and processed is virtually unlimited. Data collection by successive 'interrogation' of the measuring positions (the strain gauges or other transducers) is so rapid, and the capacity of suitable data storage media (e.g. magnetic tape) is so great, that there is no limit to the number of transducers that the equipment can cope with.

The interrogation procedure is controlled by the computer, and the actual sequence is established by the software program which can be readily adapted to suit a particular experiment. The instant at which each measurement is performed can be specified to within fractions of a millisecond. The measuring range of the voltmeter can be chosen individually. The possibility that the voltage will exceed the range is monitored and may produce an automatic control decision.

The loading devices which apply the loads to the model are program-controlled. Instead of the time-consuming application of loads to the model, the experimenter's task consists in feeding control data into the system. The loading sequence and procedure can be modified by conditions fed back from the model itself. There is no limit to the number of loading cases that can be investigated.

Evaluation and analysis of the large numbers (sometimes running into hundreds of thousands) of measured values obtained can of course be performed only by means of the computer. Conversion of these data into information useful to the engineer can thus be accomplished in a matter of minutes. Within this period of time, the model is scrutinised many times more fully than would previously have been possible in weeks of laborious work. Finally the results of the computer output may be expressed in graphical form.

The model itself has to be constructed manually: this continues to be the distinctive feature of model analysis. With rapid modern methods of collecting and processing the data, the time taken to construct the model and test rig is virtually the only factor determining the time necessary to carry out a series of tests.

In Section 4.1 it was noted that the finite element method has a practical limit in that the load of work to be handled by the computer increases exponentially with the degree of difficulty or complexity of the problem. When functioning as a partner in hybrid analysis, i.e. operating in conjunction with a structural model which is being tested, the computer is not affected by such limiting conditions. Although it has to handle a vast number of data which represent the exact behaviour of the structure, and although any number of loading conditions can be analysed, all this does not exceed the capacity of a modern mini-computer. This is bound up with the nature of the analytical duties allocated to the computer. These consist merely in simple matrix operations such as addition and multiplication. Hardly any inversions are necessary. Hence the demands made upon the computer increase in only an approximately linear manner with increasing difficulty of the problem.

One of the first automatic model-testing installations will now be described (see Figure 4.3). This equipment has already been in practical use for some time and embodies a number of features which are characteristic of hybrid analysis, though not yet satisfying the ideal conditions. It was constructed with the object of obtaining influence functions for substantially plane structures (bridge decks). A small fixed-program electronic computer monitors and controls the experimental procedure. As viewed from the computer, this procedure is as follows:

The computer receives its instructions for controlling the loading device for a particular experiment. These data, recorded on punched tape, consist of the identification numbers of the various load application points with their associated co-ordinates x and y. The computer seeks out the data relating to the first load application point and then sets the mechanism of the loading device in motion (Section 3.2.3.3). The co-ordinate carriage, moving under constant supervision of the computer, travels into position over the first load application point, but without as yet applying any load to the model. As soon as the computer receives the message that the carriage has reached the load application point, it starts the interrogating system for 'collecting' the measured values. In the example described here the measuring positions are not individually interrogated by the computer but are scanned by a simple sequentially functioning step-by-step relay which switches from one position to the next. The measuring signals corresponding to this zero loading are amplified by means of a carrier-frequency measuring bridge and then converted into digital form by an analog–digital converter. The digital values of a measuring sequence (in this instance comprising up to 40 measuring positions) are fed into the computer and temporarily stored in its core. The load is now applied at the point concerned and the measured values are again transmitted

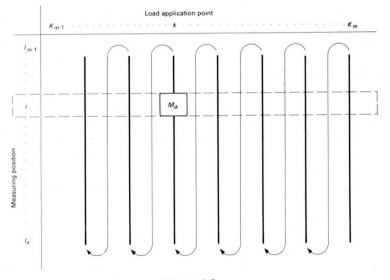

Figure 4.3

in the same sequence to the central unit. As it receives each measured value, the computer calculates the difference between it and the stored zero value; this difference is then passed to two recording devices, namely a printer and a card punch. On completion of the recording operation, a signal issued by this peripheral equipment initiates the reception of the next measured value. The speed with which the successive measuring positions can be interrogated is thus limited by the slowest device in the system—the card punch—to about one measurement per second. The interrogation of the zero values is performed before each new load application. The object of this is twofold. Firstly, it is no longer necessary to establish a null balance for each measuring position before measurement is carried out; secondly, the risk of error arising from a possible zero drift of the amplifier, which could occur during measurements extending over a relatively long period of time, is entirely eliminated. The computer includes a logic system whereby the measured values can, by selection of a multiplication factor, be multiplied by a constant while the experiment is in progress.

On the paper tape on which the output from the printer is recorded the measured values are presented in a form identical with that on the automatically-punched cards. The tape provides a means of continuously checking the values obtained and thus monitoring the experimental procedure. For the further processing of the measured values (see Sections 4.2.2 and 4.2.3) only the punched cards are utilised. These are fed into the computer provided with the appropriate evaluating and analysing programs.

Programing the punched tape to control the application of the load is not automatic. The loading device itself serves as the input apparatus for this purpose. The load application pin can be removed and replaced by an optical projection device which forms the image of a cross-wires on the model at the intended point of application of the load. The co-ordinate carriage is guided to the required load application point by manual control, and when a push-button is operated, the co-ordinates x and y, together with the identification number of the load application point, are recorded by the tape punch. In this way, the whole loading program takes only a few minutes to complete.

The apparatus described above is perhaps of some historical value in that it was the first experimental installation in which the loading and measuring sequence was fully automated and computer-controlled. It still falls short of the ideal conditions required for modern hybrid analysis however. More particularly, it lacks the flexibility of a computer as the central controller of the sequence of operations and it also lacks the capability of directly analysing the measured data.

A new and more versatile installation (Fig. 4.4) is under development. Its central feature is a rapid process computer with the measured data being recorded with a frequency of approximately 30 channels per second by means of an integrating voltmeter direct on to magnetic tape. The computer is thus able to perform all the analytical operations for model testing.

The operational flexibility and thus the value of the whole apparatus is substantially enhanced by a graphic input and output device specially

Figure 4.4   Testing installation for hybrid analysis (see also Figures 3.26 and 3.45).

developed for application to model testing. This apparatus has been evolved with the thought that any new method of structural analysis associated with the use of a computer is proportionately of greater value to the design engineer if he can simplify the form of his input data and if the results are presented to him in a convenient visual form (graphic output). The following four distinctive features of this equipment are designed to serve this purpose:

(i) First, the apparatus is a plotting machine which, under the control of the computer, is able to present the results (e.g. principal stresses, deflected surfaces, etc.) in the form of multicoloured diagrams with captions and lettering.

(ii) Input of co-ordinates and angles is possible by means of an optical positioning device (similar to that on the loading apparatus described earlier) whereby, for example, the geometric features of models and the nature and location of measuring positions can be quickly and conveniently recorded.

(iii) A keyboard is available for feeding numerical information into the computer. Thus, combinations of figures can be used to identify points (e.g. recorded optically), designate the addresses of sub-routines in a computer and supply these programs with data.

(iv) By means of a photo-electric tracing device, the apparatus is able to read curves. Thus the engineer can, for example, record the pattern of a load (or the profile of a pre-stressing cable). The reading device then automatically informs the computer of the geometric shape of this input quantity.

4.2.2 INFLUENCE FUNCTIONS AND HYBRID ANALYSIS

4.2.2.1 Some comments on structural analysis methods

In the design and analysis of framed structures the engineer will, as a matter of course, make use of influence lines for determining the limiting values of the stresses, stress resultants, deformations, etc. that arise under live load. On the other hand, in seeking to solve problems associated with plate and shell structures, the designer can make use of influence surfaces only in those relatively rare cases where these surfaces can be determined from tabulated numerical data compiled for the purpose. The evaluation of influence surfaces is laborious and restricted to a few selected stress resultants and types of loading. In most cases we must content ourselves with estimating the load limits of such two- or three-dimensional structures from a few individual load patterns or combinations, the choice of which is left to the personal judgment of the designer. A straightforward general procedure, such as we possess for the analysis of framed structures, has hitherto not been available for plates and shells. We have evidently become accustomed to accepting this qualitative difference in the methods of analysis for these two major classes of structure.

Even the application of large computers to the solution of these problems will not significantly change this state of affairs in the near future. Because of limitations of capacity to perform elaborate matrix inversions, the digital computer already has trouble in analysing individual loading cases on plate and shell structures of relatively modest complexity.

4.2.2.2 Influence functions

Despite the great advantages that the knowledge of a large number of influence surfaces offers the structural designer, present-day computer programs must for economic reasons dispense with the systematic calculation of influence functions even in the relatively simple cases where this is theoretically feasible. Simply in terms of computational effort, a solution procedure involving influence functions constitutes an unacceptable detour for a limited number of individual problems.

From the latest developments in the field of model analysis and testing, on the other hand, it emerges very clearly that with present-day measuring and control techniques any number of influence functions can quickly be obtained from an elastic structure, no matter how complex it is. It is also apparent that this procedure based on the use of influence functions, which in terms of digital computing is a troublesome detour, is a most natural and economical procedure in the field of model analysis. In combination with the computer functioning as a 'process computer', it is possible with the new model-testing techniques to produce automatically various types and conditions of loading and to measure the effects quickly and accurately. The model can thus be made to yield a vast amount of relevant information which is at once interpreted into the language of the structural engineer.

With modern technical resources it takes only a few hours to perform the experimental determination of hundreds of thousands of influence coefficients.

When the influence functions are known it is possible not only to analyse any loading case and systematically determine limit values but also to simulate altered boundary conditions, or modifications to the shape of the elastic structure under investigation, without having to go back to the model itself.

This modern approach and technique, which differ from conventional model analysis in some very significant respects, are referred to as 'hybrid analysis' to symbolise the close relationship between the model (the analog system) and the digital computer functioning as controller and interpreter.

The actual technique of hybrid analysis is outside the scope of this book. The information given in the following sections merely aims at giving the reader a glimpse of the many simple, elegant and hitherto little known possibilities available to the design engineer using this novel technique.

4.2.2.3 Influence coefficients, influence functions

Consider a homogeneous isotropic elastic structure K. Its state of strain and hence its state of stress under known loading are determined by the material properties (modulus of elasticity, Poisson's ratio), the geometrical shape and the type of support (boundary conditions). At the point $l(x, y, z)$ we apply a force and moment vector (an action F, M) and at the point $j(u, v, w)$ we determine the displacement in the form of a translation and rotation vector (an effect D, R) (Fig. 4.5). The points j and l are accordingly referred to as the effect point and the action point respectively.

Actions and displacements each possess six degrees of freedom since each of their two vectors is determined by three components. As our structure is assumed to be elastic, if we consider deformations of small magnitude the components are linked by the following linear relationships for all pairs of points j, l:

$$D_u = \alpha_{DFux}F_x + \alpha_{DFuy}F_y + \alpha_{DFuz}F_z + \alpha_{DMux}M_x + \alpha_{DMuy}M_y + \alpha_{DMuz}M_z$$

$$D_v = \alpha_{DFvx}F_x + \alpha_{DFvy}F_y + \alpha_{DFvz}F_z + \alpha_{DMvx}M_x + \alpha_{DMvy}M_y + \alpha_{DMvz}M_z$$

$$D_w = \alpha_{DFwx}F_x + \alpha_{DFwy}F_y + \alpha_{DFwz}F_z + \alpha_{DMwx}M_x + \alpha_{DMwy}M_y + \alpha_{DMwz}M_z$$

$$R_u = \alpha_{RFux}F_x + \alpha_{RFuy}F_y + \alpha_{RFuz}F_z + \alpha_{RMux}M_x + \alpha_{RMuy}M_y + \alpha_{RMuz}M_z$$

$$R_v = \alpha_{RFvx}F_x + \alpha_{RFvy}F_y + \alpha_{RFvz}F_z + \alpha_{RMvx}M_x + \alpha_{RMvy}M_y + \alpha_{RMvz}M_z$$

$$R_w = \alpha_{RFwx}F_x + \alpha_{RFwy}F_y + \alpha_{RFwz}F_z + \alpha_{RMwx}M_x + \alpha_{RMwy}M_y + \alpha_{RMwz}M_z.$$

$$(4.1)$$

The coefficients α_{pqik} are referred to as 'influence coefficients'. They represent the ratio of the effect to the action. From equation (4.1), it appears that for the complete description of the elastic reciprocal relation between a displacement and the action that produces it, we require 36 influence coefficients for

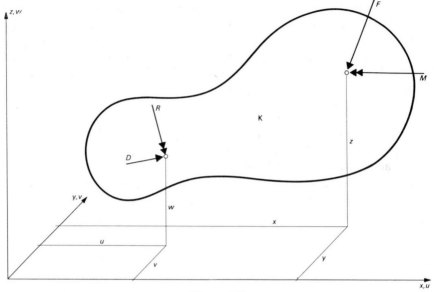

Figure 4.5

each pair of points. The relationship given here is an example of a multiplicity of reciprocal influences between conceivable actions and effects. Inversion of the influence matrix is equivalent to interchanging the action and the effect and is physically equally meaningful. There are of course influence coefficients for other quantities, such as stress resultants, stresses, strains, etc.

We can distinguish between various types of influence coefficients. They are characterised by the action/effect pair associated with each of them; in equation (4.1) this pair is designated by the first two subscript letters of the coefficient α. Hence an influence coefficient can, in general, be characterised as follows:

$$P_i = \alpha_{p,q,i,k} Q_{jk} \qquad (4.2)$$

where p stands for the effect and q for the action type of influence coefficient. The subscripts i and j refer to the appropriate components of the vectors P and Q. They are associated individually or groupwise with the geometric points j and l.

The dimension of α is that of P/Q. In model analysis it is especially important that this dimension be carried through the calculation, since the dimension, together with the scale of the model and the modulus of elasticity, determines the laws of similitude for the conversion of 'model quantities' to actual physical quantities in the prototype.

From the set of equations expressed in (4.1), it is possible to select the relation between the displacement vector D and the force vector F and represent them

in matrix form:

$$D = \begin{Bmatrix} D_u \\ D_v \\ D_w \end{Bmatrix} = \begin{bmatrix} \alpha_{ux} & \alpha_{uy} & \alpha_{uz} \\ \alpha_{vx} & \alpha_{vy} & \alpha_{vz} \\ \alpha_{wx} & \alpha_{wy} & \alpha_{wz} \end{bmatrix} \begin{Bmatrix} F_x \\ F_y \\ F_z \end{Bmatrix} = [\alpha_{ik}]F. \tag{4.3}$$

To determine the displacement vector at any point j caused by the application of a force vector at a point l, it is therefore necessary to know nine influence coefficients α_{ik}. Using the general notation P to denote the effect vector and Q to denote the action vector, we can write:

$$P = [\alpha_{ik}]Q. \tag{4.4}$$

The contents of the matrix $[\alpha_{ik}]$ can now be interpreted geometrically. The three elements of each row in the influence matrix (4.3) can be imagined as being the components of a vector, the 'influence vector'. Each component of the effect vector is calculated by scalar multiplication of the action vector by the corresponding influence vector:

$$P_u = \vec{\alpha}_u Q$$
$$P_v = \vec{\alpha}_v Q \tag{4.5}$$
$$P_w = \vec{\alpha}_w Q.$$

We shall now suppose that the effect point $j(u_0, v_0, w_0)$ is stationary and see what happens to the effect P_0 when we allow the action point to move about on the structure with its direction and magnitude varying.

In order to describe this behaviour it is obviously necessary to know the spatial functions of $[\alpha_{ik}]$ and Q. We can therefore write:

$$P_{u_0} = \vec{\alpha}_{u_0}(x, y, z)Q(x, y, z)$$
$$P_{v_0} = \vec{\alpha}_{v_0}(x, y, z)Q(x, y, z) \tag{4.6}$$
$$P_{w_0} = \vec{\alpha}_{w_0}(x, y, z)Q(x, y, z).$$

The components of $\vec{\alpha}$ are referred to as 'influence functions'. Three of these influence functions together define an 'influence vector field'. One such (potential) vector field is associated with each directionally-determined effect. At each point, it provides the multiplier for determining a real effect caused by action vectors acting somewhere on the structure.

We shall now—quite abstractly, to begin with—establish the curl of one of these vector fields $\vec{\alpha}$, i.e. by means of this operation, we shall establish a new vector field $\vec{\beta}$:

$$\vec{\beta} = \text{curl } \vec{\alpha}. \tag{4.7}$$

Its components are:

$$\beta_x = \frac{\partial \alpha_z}{\partial y} - \frac{\partial \alpha_y}{\partial z} \qquad \beta_y = \frac{\partial \alpha_x}{\partial z} - \frac{\partial \alpha_z}{\partial x} \qquad \beta_z = \frac{\partial \alpha_y}{\partial x} - \frac{\partial \alpha_x}{\partial y}. \tag{4.8}$$

This newly-obtained vector field is important in hybrid analysis as will become apparent in due course (Fig. 4.6). To enable us to visualise this, let us imagine that the action vector is a force F acting at the point x, y, z. We shall now investigate the influence exercised by each of its components independently; to start with, we shall consider the two components located parallel to the x–y plane. Displacement of the component F_x by an amount Δz and displacement of F_z by an amount Δx produces a moment $\Delta M_y = F_x \Delta z - F_z \Delta x = F \Delta s$.

Since the moment can be produced by a displacement which is perpendicular to the force F and acts in any direction $\varphi + \pi/4$, it is permissible to allow F to pass through the two points of application Δx and Δz, respectively, of the force components. In addition the magnitudes of F_x and F_y can be chosen at random, provided that the sum of the moments produced by their displacement is equal to ΔM. We can therefore write:

$$\Delta z = \frac{1}{2}\frac{\Delta s}{\cos \varphi} \qquad\qquad F_x = F \cos \varphi$$

$$\text{and}$$

$$\Delta x = \frac{1}{2}\frac{\Delta s}{\sin \varphi} \qquad\qquad F_z = F \sin \varphi .$$

From the diagram, it may be deduced that the change of the effect:

$$\Delta P = F_x \frac{\partial \alpha_y}{\partial z}\Delta z + F_z \frac{\partial \alpha_z}{\partial x}\Delta x.$$

On substituting the above relations, we obtain the following expression for the influence coefficient of a moment:

$$\frac{\Delta P}{\Delta M_y} = \frac{1}{2}\left(\frac{\partial \alpha_x}{\partial z} - \frac{\partial \alpha_z}{\partial x}\right) = \frac{1}{2}\beta_y .$$

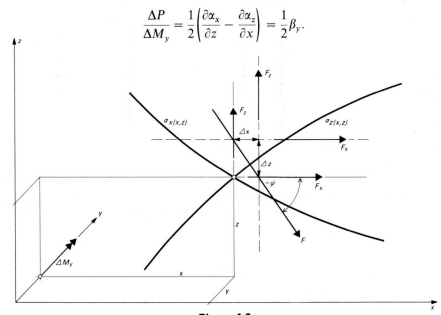

Figure 4.6

By cyclic transposition of the subscripts we obtain the influence of all the moment components in accordance with (4.8).

If $\vec{\alpha}$ is the influence vector field for forces, then the influence field for moments acting on the same elastic structure K is represented by $\frac{1}{2}$ curl $\vec{\alpha}$.

We can therefore replace β by the influence coefficients with the appropriate subscripts and write, for example:

$$\vec{\alpha}_M = \tfrac{1}{2} \operatorname{curl} \vec{\alpha}_F. \tag{4.9}$$

A similar relation exists between the influences exerted by the translation and rotation actions:

$$\vec{\alpha}_R = \tfrac{1}{2} \operatorname{curl} \vec{\alpha}_D. \tag{4.10}$$

We have thus established that the influence functions associated with the various action types for one elastic system are analytically interconnected. The prerequisite condition is that the functions must be continuous and differentiable. Adopting the notation:

$$\operatorname{curl}[\alpha] = \begin{bmatrix} \operatorname{curl} \vec{\alpha}_u \\ \operatorname{curl} \vec{\alpha}_v \\ \operatorname{curl} \vec{\alpha}_w \end{bmatrix} \tag{4.11}$$

we can write the first three lines of (4.1) in the following abridged form:

$$D = [\alpha_{DF}]F + \tfrac{1}{2} \operatorname{curl}[\alpha_{DF}]M. \tag{4.12}$$

We have thus obtained a relationship in which the number of unknown influence functions is reduced by half in comparison with the number of required influence coefficients. For hybrid analysis this means that influence functions for moment and for rotation can be calculated analytically from the experimentally-determined influence functions for force and displacement, respectively.

If we now also allow the effect point $j(u, v, w)$ to be a variable, it is possible to obtain the most general form of influence function:

$$[\alpha_{ik}(u, v, w, x, y, z)]. \tag{4.13}$$

The analytical relationships for the effects are of course similar to those given in (4.9) and (4.10), except that now the curl operation must be performed between the columns of the influence matrix which are conceived as components of vectors.

Let the curl thus defined for the effect points be represented by curl_j (derivatives with respect to u, v and w) and let curl_l represent the curl at the action points. Then the set of equations expressed in (4.1) can be written entirely in terms of the influence relationships between force and displacement:

$$D = [\alpha]F + \tfrac{1}{2} \operatorname{curl}_l[\alpha] \cdot M$$
$$R = \tfrac{1}{2} \operatorname{curl}_j[\alpha] \cdot F + \tfrac{1}{4} \operatorname{curl}_j(\operatorname{curl}_l[\alpha])M. \tag{4.14}$$

Each of the matrices indicated is a square matrix comprising nine elements

(functions) whose three rows represent the action-dependent vectors and whose columns represent the effect-dependent vectors. The matrix $[\alpha]$ contains the elements indicated in equation (4.3).

The first row of the curl matrix $\text{curl}_v [\alpha]$ consists of the following components:

$$\frac{\partial \alpha_{uz}}{\partial y} - \frac{\partial \alpha_{uy}}{\partial z}; \quad \frac{\partial \alpha_{ux}}{\partial z} - \frac{\partial \alpha_{uz}}{\partial x}; \quad \frac{\partial \alpha_{uy}}{\partial x} - \frac{\partial \alpha_{ux}}{\partial y}.$$

The first column of $\text{curl}_j [\alpha]$ is:

$$\frac{\partial \alpha_{wx}}{\partial v} - \frac{\partial \alpha_{vx}}{\partial w}$$

$$\frac{\partial \alpha_{ux}}{\partial w} - \frac{\partial \alpha_{wx}}{\partial u}$$

$$\frac{\partial \alpha_{vx}}{\partial u} - \frac{\partial \alpha_{ux}}{\partial v}.$$

The general force–displacement relationship:

$$D = [\alpha_{ik}]F \tag{4.15}$$

describes the elastic behaviour of the structure completely. By means of further operations using known relationships, the elastic reciprocity between all the physical quantities commonly encountered in structural analysis (such as stress and strain, internal moments, restraints, etc.) can be expressed.

The great disadvantage of this procedure for solving problems with the aid of influence functions is its complexity. Even for relatively simple plate and shell structures these functions are seldom determined and even with modern computer methods they can be evaluated only in exceptional cases.

However, with hybrid analysis it is possible to determine experimentally, and with comparatively little effort, a sufficient number of exact influence coefficients which can then be treated further as functions with the aid of the computer.

In practical cases it is unnecessary to know the relationships between all the components at all the action and effect points. For example, consider a structure located substantially in the horizontal $(x–y)$ plane (in the ideal case this would be a slab, but most bridge decks can be regarded as coming within this category). Let us assume, to start with, that it is subject only to vertical loading. The influence function for each effect point P will then have the following form:

$$P = \alpha_{Fz}(x, y)F_z. \tag{4.16}$$

Let $\alpha_{Fz}(x, y)$ be the influence surface for P due to a moving vertical concentrated load F_z.

With the aid of the relationship (4.9), we obtain the influence surface for P due to moving moments:

$$\alpha_{Mx} = \frac{\partial \alpha_{Fz}}{\partial y} \tag{4.17}$$

$$\alpha_{My} = -\frac{\partial \alpha_{Fz}}{\partial x}. \tag{4.18}$$

As an example we will investigate, with the aid of the experimentally-established function (4.16) and the functions (4.17) and (4.18), the effect of a pre-stressing cable on the associated effect point $P(u_0, v_0)$ (Fig. 4.7).

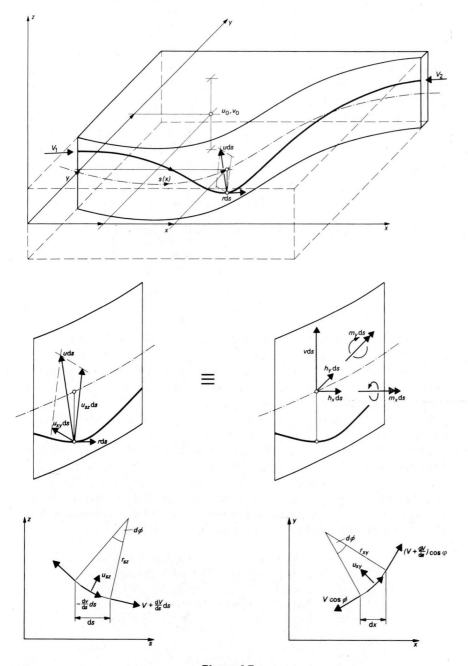

Figure 4.7

The geometric location of the cable is determined by the vertical co-ordinate:

$$z = f(s).$$

and the profile on plan:

$$s = g(x).$$

The forces exerted by the cable upon the structure comprise the anchorage forces V_1 and V_2, the 'radial' forces $u(s)$ (these are exerted on the concave side of the cable where the latter is curved and act at right angles to the centre-line of the cable) and the frictional forces $r(s)$.

Let the variation of the pre-stressing force along the cable be defined by the function $V = V(s)$. This is determined by a boundary condition and the (assumed) relationship for the friction. The distribution of the friction along the cable is then:

$$r(s) = -\frac{dV}{ds}.$$

The two components of u in the s–z and x–y planes, respectively, are:

$$u_{sz}(s) = \frac{V(s)}{r_{sz}(s)}$$

$$u_{xy}(s) = \frac{V(s) \cos \varphi}{r_{xy}(s)}$$

where φ denotes the slope of the centre-line of the cable with respect to the x–y plane.

The force vectors $u\,ds$ and $r\,ds$ can now be transformed into an action at the neutral axis, consisting of the force components $v\,ds$, $h_x\,ds$ and $h_y\,ds$ and the moment components $m_x\,ds$ and $m_y\,ds$.

The horizontal components produce a purely two-dimensional state of stress. The effect of this will be considered later.

Apart from the normal force in the plane of the plate, the effect P can now be calculated as follows by means of the influence function $\alpha_{F_z}(x, y)$:

First, we apply a vertical cut through the influence surface and thus deter-mine the influence line $\alpha(s)$ along the cable profile on plan (projected curve s). We shall also express the partial derivatives of α_{Mx} and α_{My} as functions of s. The overall influence of the pre-stressing cable on P is then:

$$P = V_1 \sin \varphi_1 \cdot \alpha(s_1) + \int_1^2 \alpha(s) \cdot v(s)\,ds + V_2 \sin \varphi_2 \cdot \alpha(s_2)$$

$$+ V_{1x} \cos \varphi_1 \frac{\partial \alpha(s_1)}{\partial y} + \int_1^2 \frac{\partial \alpha(s)}{\partial y} m_x(s)\,ds + V_{2x} \cos \varphi_2 \frac{\partial \alpha(s_2)}{\partial y}$$

$$- V_{1y} \cos \varphi_1 \frac{\partial \alpha(s_1)}{\partial x} - \int_1^2 \frac{\partial \alpha(s)}{\partial x} m_y(s)\,ds + V_{2y} \cos \varphi_2 \frac{\partial \alpha(s_2)}{\partial x}. \quad (4.19)$$

At first sight, this method of calculating the effect of the pre-stress on a single effect point P may appear very complicated. But if we consider that the operation is performed by the computer within fractions of a second we can appreciate the great advantage achieved by the application of influence functions to the analysis of the pre-stress.

Any number of variants of this loading case can be analysed simply by altering the geometry of the cable profile and the optimum solution thus found. For plate and shell structures this would be impracticable with any of the other known methods of structural analysis.

The expression (4.19) calls for the following further comment. The influence of the normal forces associated with the pre-stress, namely, $V_1 \cos \varphi_1$ and $V_2 \cos \varphi_2$, and the influence of h_x and h_y have been neglected. Nevertheless, P represents the exact solution, since these neglected quantities produce a state of pure membrane stress in the plate (or slab) under consideration, or if P is for example a reaction, a vertical deflection or a moment in the plate. If the edge stresses (extreme fibre stresses) are also of interest, then the influence functions

$$\alpha_{Fx}(x, y) \quad \text{and} \quad \alpha_{Fy}(x, y)$$

must also be known, i.e. they will have to be determined experimentally or (possibly) be introduced analytically.

4.2.2.4   Influence matrices

The important basic parameters for hybrid analysis, the influence functions, cannot be determined experimentally as ideal continuous functions, nor can they be established as such by means of the computer. Instead, the hybrid investigation technique must consist of producing a sufficiently large number of influence coefficients for the analytical operations to be performed numerically with satisfactory accuracy.

For practical calculations it is very important to produce actions at the largest possible number of points in order that the actual loading cases to be introduced later can be freely chosen. The directly measured effects on the other hand should be confined to the points of interest. It is most important for the evaluation and assessment of the results of measurements, however, that all the bearing reactions of the structure be determined.

The process of establishing a basic matrix which describes the behaviour of an elastic structure in the most general form is undertaken in two steps:

First, the direct D matrix is produced. This is formed automatically and on-line with the computer, which controls the application of the actions, measures the effects and (before storing the data) interprets, normalises and converts the measured values. The number of elements in that matrix therefore corresponds to the number of measurements performed, i.e. to the product of the number of action points and the number of effect points where measurements are obtained. A D matrix is represented in Figure 4.8. Each column corresponds to an action point k and each row to an effect point i. During the

Direct matrix

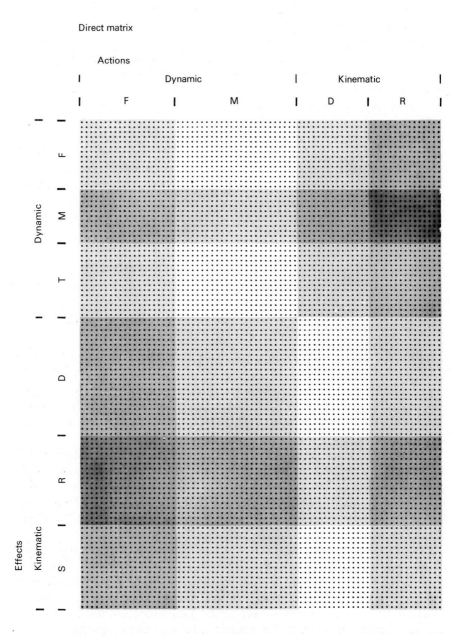

Figure 4.8

Model matrix

System matrix

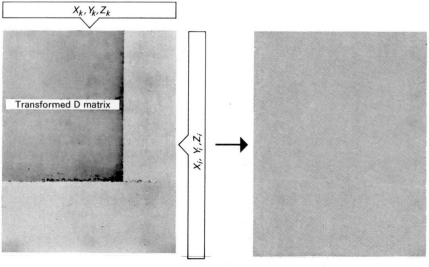

Figure 4.9

execution of the test all the effects are measured for each action, i.e. the matrix is filled up column by column.

During the measuring operation the computer converts the measured values m_{ik} into influence coefficients α_{ik}. In doing this account is taken of the following principles: (a) the reference values of the electrical measuring transducers supply the mechanical quantities; (b) the measured values are converted to influence coefficients by taking due account of the actions applied; (c) the effects on the model are transformed into effects on the prototype by means of a similarity conversion.

The matrix in Figure 4.8 is split up into sub-matrices which characterise the type of influence coefficients they contain and hence their dimensions. Each dimension corresponds to particular model similitude laws which are represented in the corresponding compartments of the matrix.

In the second step, the D matrix is expanded into an M matrix (Fig. 4.9) by numerical operations in the computer. First, the geometric locations of all the effect points and action points are established by introducing a co-ordinate system. With this information the computer can treat certain groups of influence coefficients approximately as influence functions. For instance, if the force actions denote vertical concentrated forces applied at various points, the associated set of rows for each effect will represent an influence surface as in Figure 4.10. Any required influence coefficients on the surface can now be determined by interpolation. More particularly, influence lines along a curve defined in plan (i.e. in projection on the horizontal plane) can be calculated. Such curves may, for example, represent traffic lanes on a bridge deck, railway tracks, or profiles of pre-stressing cables. As can readily be seen, such

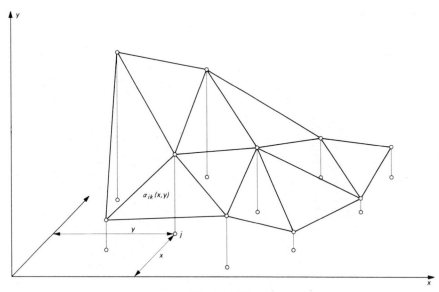

Figure 4.10

influence lines can be evaluated as easily as the influence lines encountered in the analysis of framed structures.

We may obtain influence coefficients for new effects by differentiation. In Figure 4.11 some measured influence coefficients (a section through an influence surface due to vertical concentrated loads) have been plotted as an influence line, and its derivative function is also plotted. We see that if a high-precision experimental technique is employed, the sequence of influence coefficients obtained can be treated as a continuous function.

In addition to new, indirectly measured actions, it is also analytically possible to obtain influence coefficients for other effects. In particular, the experimental research engineer will want to calculate influence coefficients for those effects which will be of interest to him during the subsequent determination of limit values.

Some examples are as follows:

(i) Strain measurements obtained with rosettes may be transformed into stresses in prescribed directions.

(ii) Bearing reactions may be represented by components in the direction of the co-ordinate axes.

(iii) Influence functions of overall stress resultants may be obtained at any desired section from equilibrium calculations.

(iv) The pattern of local stress resultants may be determined from sets of strain measurements.

Expressed in the language of finite elements, the M matrix is simply a huge combined stiffness and deformation matrix for a highly complex element which in general (but not necessarily) comprises the whole structure under

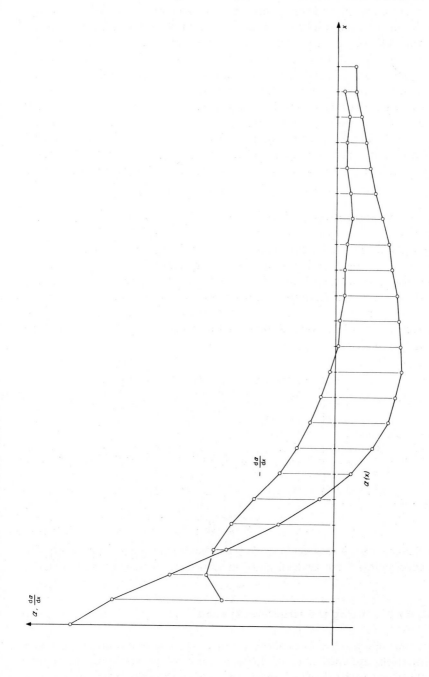

Figure 4.11

investigation. Hence the elastic behaviour of the structure (as contrasted with the finite element) may be exactly described, since the model forces and displacements have resulted from a realistic strain or stress function in the element. Of course, these elastic 'macro-elements' can (just like the finite elements) be joined together analytically to form new structures or, alternatively, they can be introduced as components or sub-assemblies during the computer analysis of a larger elastic structure.

With all the attractive possibilities opened up by hybrid analysis, it must nevertheless be borne in mind that in order to obtain reliable influence coefficients from the analogue system (i.e. the structural model) it is essential to use an experimental technique which as far as precision, adaptability and speed is concerned is superior to the conventional technique of model testing.

4.2.2.5 Superposition

If the influence functions are known, calculating the effects due to any particular loading case is an easy task for the computer. The solution of the difficult elastic problem is already contained in these functions. The results for all the effects are obtained by applying load factors to all the load points k, multiplying them by the associated influence coefficients and summing these products at the effect point. Represented as a matrix operation, this can be written as follows:

$$
\begin{Bmatrix} P_1 \\ P_2 \\ \vdots \\ P_i \\ \vdots \end{Bmatrix} = \begin{bmatrix} \alpha_{11} & \alpha_{12} & \cdots & \alpha_{1k} & \cdots \\ \alpha_{21} & & & \vdots & \\ \vdots & & & \vdots & \\ \alpha_{i1} & \cdots\cdots\cdots\cdots & \alpha_{ik} & \vdots \\ \vdots & & & \vdots & \end{bmatrix} \begin{Bmatrix} Q_1 \\ Q_2 \\ \vdots \\ Q_k \\ \vdots \end{Bmatrix}
$$

or simply:

$$P = [\alpha_{ik}]Q.$$

Of course, any desired set of loads Q can be defined and the limit values for the effect points P can be determined in accordance with any desired criteria.

4.2.2.6 Modifying the structural system

In the superposition procedure, external actions, whose magnitudes and distributions are known, act as 'loading cases' on the structure. The results, i.e. the effects thus calculated, reflect the behaviour of the elastic structure as described in the M matrix, with due regard to the boundary conditions mechanically realised in the model.

We now wish to transform the M matrix by purely analytical means into a new matrix (the T matrix) which represents the elastic behaviour of our structure under altered boundary conditions. In this way we simulate a structural system differing from that of the model, and we do this without the need for fresh experimental data.

First of all we shall draw a distinction between: (a) active actions, namely known loading cases such as external loads which are applied to the structure on the superposition principle; and (b) passive actions, namely selected actions located outside the boundary of the structure (or portion of a structure) under investigation and whose magnitude is so calculated that boundary effects associated with them will, under every active loading case, satisfy certain conditions.

We shall accordingly select a set of influence coefficients α_{ik} which we shall call 'indicators' and designate by the notation β_{ik} for the sake of distinction. The corresponding edge effects and edge actions will be conformably designated. The effects thus distinguished comprise a proportion of the influence exercised by the (as yet unknown) actions:

$$P_\beta = [\beta_{ik}]_n Q_\beta. \tag{4.20}$$

In contrast with $[\alpha_{ik}]$, the matrix $[\beta_{ik}]$ is always a square matrix of nth rank, since we bring each action point into relation with an effect point. If Q_β is known, the result for all effects P_i under a particular external loading can be written as follows:

$$P_i = \sum_{k=1}^{k=m} \alpha_{ik} Q_{\beta k} + \sum_{k=m+1}^{k=m+1+n} \alpha_{ik} Q_{\beta k} \tag{4.21}$$

and, more particularly, the n effects in the boundary zone under consideration:

$$\begin{Bmatrix} P_1 \\ P_2 \\ \vdots \\ P_n \end{Bmatrix} = \begin{bmatrix} \alpha_{11} & \alpha_{12} & \cdots & \alpha_{1m} \\ \alpha_{21} & & & \\ \vdots & & & \\ \alpha_{n1} & & & \alpha_{nm} \end{bmatrix} \begin{Bmatrix} Q_1 \\ Q_2 \\ \vdots \\ Q_m \end{Bmatrix} + [\beta_{ik}]_n \{Q_\beta\}_n.$$

Let O denote the proportion of the effect due to the m active actions. We can then write:

$$P_\beta = P_{\beta 0} + [\beta_{ik}] Q_\beta. \tag{4.22}$$

We shall now use this relationship to formulate general boundary conditions by establishing the requirement that the following relationship must exist between the actions and their associated effects:

$$P_\beta = \begin{Bmatrix} c_1 & Q_1 \\ c_2 & Q_2 \\ \vdots & \\ c_n & Q_n \end{Bmatrix} = \{cQ_\beta\}.$$

The scalar factors c have the significance of general 'spring constants'. They are a measure of the stiffness of the bearings and we are free to choose their magnitude.

On introducing the relationship

$$
\begin{Bmatrix} c_1 & Q_1 \\ c_2 & Q_2 \\ \vdots \\ c_n & Q_n \end{Bmatrix} =
\begin{bmatrix} c_1 & 0 & \cdots & 0 \\ 0 & c_2 & & \vdots \\ \vdots & & \ddots & \vdots \\ 0 & \cdots\cdots & & c_n \end{bmatrix} \{Q_\beta\}
$$

into (4.22), we obtain:

$$
0 = ([\beta] - [c])Q_\beta + P_{\beta 0}
$$

which is a set of linear equations that we can solve for Q:

$$
\begin{Bmatrix} Q_{\beta 1} \\ Q_{\beta 2} \\ \vdots \\ Q_{\beta n} \end{Bmatrix} = -
\begin{bmatrix} (\beta_{11} - c_1) & \beta_{12} & \cdots & \beta_{1n} \\ \beta_{21} & (\beta_{22} - c_2) & & \vdots \\ \vdots & & \ddots & \vdots \\ \beta_{n1} & \cdots\cdots & & (\beta_{nn} - c_n) \end{bmatrix}^{-1}
\begin{Bmatrix} P_{10} \\ P_{20} \\ \vdots \\ P_{n0} \end{Bmatrix} . \tag{4.23}
$$

On substituting the result for Q_β into (4.21), we obtain all the effects relating to the modified structural system for a loading case defined by Q_k.

This transformation may also be performed direct on the M matrix without assuming an actual loading case. Each column of the matrix represents the effects for one unit loading case. Let α_{ik}^T denote the influence coefficients of the new (transformed) matrix; then the following relationship holds:

$$
\alpha_{ik}^T = \alpha_{ik} + \sum_{k=m+1}^{k=m+1+n} \alpha_{ik} q_{\beta k} . \tag{4.24}
$$

The passive action factors q_{ik} have been written with lower case letters in this expression in order to indicate that we are dealing here not with actual loads but with dimensionless quantities.

As may be seen, the calculated passive actions act as correction factors which convert the effects measured on the model serving as the structural 'basic system' into effects on a structure with altered boundary conditions. This process will be illustrated by the following examples.

4.2.2.7   Examples

Removal of supports

Suppose that we have established the influence matrix from measurements performed on the structural model of a bridge. We wish to investigate how the structure will behave for the loading case of dead weight g when three supports, which in the model consist of vertical rocker columns, are removed. It will

be assumed that the influence-enforced vertical displacements D of the supports are contained in the M matrix.

The results for reactions, deflections and strain in the original system are obtained from superposition with the load factors Q_k associated with the load positions:

$$P_{0i} = [\alpha_{ik}]Q_k.$$

The bearing reactions at the supports which we intend to remove are contained in the result P_0. Let these be:

$$\begin{Bmatrix} F_{01} \\ F_{02} \\ F_{03} \end{Bmatrix}.$$

We now seek to calculate the displacements D_1, D_2 and D_3 which will produce at the three respective supports an opposite reaction to the original bearing reaction. On applying these displacements we cause the three bearing reactions to disappear and thus in fact eliminate the supports themselves (under the given loading case). The spring constants c_1, c_2, c_3 are therefore zero.

From the known influence coefficients of the three bearing reactions produced by three vertical unit displacements of the supports we can with the aid of (4.23) calculate the three displacements:

$$\begin{Bmatrix} D_1 \\ D_2 \\ D_3 \end{Bmatrix} = - \begin{bmatrix} \beta_{11} & \beta_{12} & \beta_{13} \\ \beta_{21} & \beta_{22} & \beta_{23} \\ \beta_{31} & \beta_{32} & \beta_{33} \end{bmatrix}^{-1} \begin{Bmatrix} F_{01} \\ F_{02} \\ F_{03} \end{Bmatrix}.$$

With these values, we can now express the results for the loadings on the new structural system:

$$P_1^T = P_{01} + \alpha_{11}D_1 + \alpha_{12}D_2 + \alpha_{13}D_3$$
$$P_2^T = P_{02} + \alpha_{21}D_1 + \alpha_{22}D_2 + \alpha_{23}D_3$$
$$\vdots$$
$$P_i^T = P_{0i} + \alpha_{i1}D_1 + \alpha_{i2}D_2 + \alpha_{i3}D_3.$$

It should be noted that the second subscript does not refer to the numbering of the original M matrix but to that of the three selected loading cases D_1, D_2 and D_3.

If we intend to calculate limit values for loading cases on the newly created structural system, it is necessary not only to transform the results for the selected loading case g (dead load) as in the above example, but also to transform the whole matrix $[\alpha_{ik}]$. The individual loading cases will be superimposed afterwards.

We have considered a support with five degrees of freedom in our example. We therefore had to know for each support the influence of one displacement quantity only in order to establish the conditions for the elimination of the reactions. If the fully restrained support is subjected to six differently orientated enforced displacements and the six associated components of the support action are measured, the effects arising from support conditions having any desired degree of freedom may be simulated.

Installing new supports

The procedure is now the opposite of that adopted in the preceding example. At the support envisaged it is necessary to know the displacements produced by the original loading case and also to know the influence coefficients due to unit forces. The displacements are equated to zero and the forces needed for satisfying this condition are calculated. Their effect is superimposed on the original matrix.

Displacement of supports

The effects arising from the displacement of supports can obviously be represented by the combination and superposition of the possibilities discussed in the two above examples. This capacity of hybrid analysis is of major practical importance since it enables the design engineer to find by simple means the optimum disposition of the supports.

Elastically deformable supports

Let us return to the first example and suppose that instead of making the bearing reactions equal to zero (as was done in that case) we consider the case where these reactions are a linear function of the settlement (vertical downward displacement) of the supports; therefore:

$$\begin{Bmatrix} F_1 \\ F_2 \\ F_3 \end{Bmatrix} = \begin{Bmatrix} c_1 & D_1 \\ c_2 & D_2 \\ c_3 & D_3 \end{Bmatrix}$$

Hence, following (4.23), we obtain for the vertical displacements at the three supports:

$$\begin{Bmatrix} D_1 \\ D_2 \\ D_3 \end{Bmatrix} = - \begin{bmatrix} (\beta_{11} - c_1) & \beta_{12} & \beta_{13} \\ \beta_{21} & (\beta_{22} - c_2) & \beta_{23} \\ \beta_{31} & \beta_{32} & (\beta_{33} - c_3) \end{bmatrix}^{-1} \begin{Bmatrix} F_{01} \\ F_{02} \\ F_{03} \end{Bmatrix}.$$

We have accordingly simulated elastic bearing conditions at the three supports, and we are free to assume any value for the spring constant for each support. In addition, it is also possible to find limit values for the effect produced by different spring constants.

In this case also the simple example may be generalised at will. If the appropriate experimental resources are available, elastic supports with varying degrees of stiffness with regard to displacement or rotation in any direction can be introduced.

Simulation of the intermediate conditions of an incomplete structure during erection

One important reason why model analysis has (with some justification) been regarded hitherto as lacking in adaptability is that once the model has been constructed its geometric features and stiffness properties are definitely fixed and can be altered only by means of material alteration of the model. Though possible, such alteration is often impracticable because of the extra time involved.

From the above examples it emerges, however, that with modern techniques the support conditions can be varied within wide limits purely by analytical simulation, without having to apply any modifications to the model at all. Theoretically the same principle applies to the geometry of the model. By adding or removing imaginary finite elements the stiffness distribution can be controlled at will, provided that at the joints or junctions between the physical model and the imaginary elements the force–deformation relationships necessary for the simulation are measured with sufficient accuracy. As far as this problem is concerned, experimental techniques are still in their infancy.

A problem which the engineer repeatedly faces is that of assessing the structural behaviour of, for example, a bridge at various stages during its construction or erection, i.e. predicting the behaviour of individual large parts or assemblies detached from the structure as a whole. As we shall see, these partial modes of behaviour can be described analytically solely on the basis of the influence functions for the whole structure, without any need to cut up the model or perform additional measurements.

Consider the bridge shown schematically in Figure 4.12. It will be assumed that the influence functions α_{ij} have been measured for the whole structure. The hatched area contains the influence coefficients of causes and effects within the region bounded by the intended cuts S—S, the whole structure being considered a continuum. We wish to find the influence functions for the part of the structure separated from the whole by the three cuts. The elastic behaviour of this particular structure can be correctly simulated if we succeed in cancelling (reducing to zero) all the stress resultants at the sections S—S for the applied loadings.

First, the stress resultants occurring at these sections must be measured on the whole system by means of an adequate number of indicators. By skilful arrangement of the measuring transducers (strain gauges) it is always possible to determine the stress resultants at a few significant positions. Since the sum of the influences at these selected positions is eventually equated to zero, only a qualitative measurement is necessary. The absolute magnitude of the stress results is of no interest. The requisite number and the disposition of the indicators will depend on the cross-section of the structure and have to be determined in each individual case. For instance, in the case of a single-cell box girder under vertical loading it is sufficient to obtain three measurements per section corresponding to the bending, torsion and shear force respectively.

For each indicator we now choose a fixed load position outside the part of the bridge deck under investigation so that large effects are produced at the

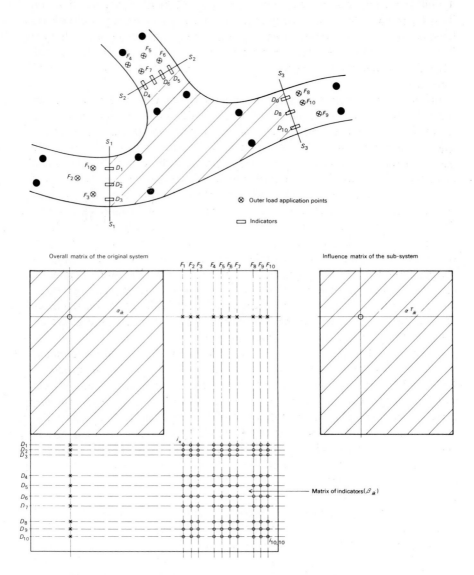

Figure 4.12   Simulation of successive stages of construction.

sections (the cuts separating that part from the rest of the bridge) and so that the values at the individual indicators are, as far as possible, independently influenced. In other words, in order to minimise the error-susceptibility of the transformation, the β matrix should, as far as possible, have a band-like character. A glance at the influence coefficients of the indicators should be sufficient to enable us to make an advantageous choice from among the possible load positions.

The new influence functions for the sub-system under consideration can now, as indicated in the drawing, be calculated in accordance with the general equation (4.24).

5

DOCUMENTATION

The following series of photographs, some showing well-known structures and the models which were tested as aids in designing them, is given intentionally without explanatory comment.

These illustrations should serve to give the reader some idea of how extensively structural model testing is utilised in helping the designer to understand and estimate the behaviour of complex and unusual structures.

5.1 Bridge over Lake Maracaibo, Venezuela

Design: Riccardo Morandi, Rome
Testing: Laboratoria Nacional de
Engenharia, Lisbon

5.2 Opera House, Sydney, Australia

Design: Jörn Utzon, Denmark
Structural Engineers: Ove Aarup, London
Testing: University of Southampton

5.3 Roof of the University Library, Basle, Switzerland

Design: O. and W. Senn, Architects, Basle
Structural analysis and testing: Author

5.4 Pirelli skyscraper, Milan, Italy

Design: Gio Ponti, Architect
Engineers: P. L. Nervi, A. Danusso
Testing: I.S.M.E.S., Bergamo

5.5 Church (Iglesia de Guadalupe), Madrid, Spain

Design Engineer: F. del Pozo
Testing: Laboratorio Central, Madrid

5.6 Structural system for Eurogas service stations

Design: L. Safier, Eurogas
Structural analysis and testing: Author

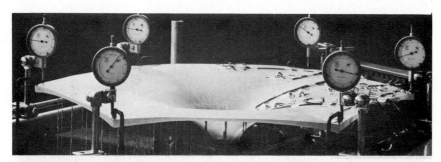

5.7 Central storage building for Verband Schweizerischer Konsumvereine, Wangen, Switzerland

Design and testing: Author

5.8 Swimming baths for the Olympic Games at Tokyo, Japan

Design: K. Tange
Engineer: Y. Tsuboi
Testing: University of
Tokyo

REFERENCES

Author	Title	Publisher/Publication
Argyris, J. H.	Recent Advances in Matrix Methods of Structural Analysis	Pergamon Press, London, 1964
Askegaard, V., Albers, A., Mortensen, P.L.	Test on a Model of Cable Roof	VDI-Berichte Nr. 102, 1966
Base, G.D.	Tests on a Perspex Model Anticlastic Roof of Lattice Construction	Cement and Concrete Association, 1962
Beaujoint, N.	Similitude et théorie des modèles	Colloque International sur les Modèles réduits de Structure, Madrid, 1959
Beaujoint, N.	Modèles réduits de résistance des caissons en béton précontraint des piles G2 et G3	RILEM Bulletin Nr. 10, 1961
Beaujoint, N., Bouché, B.	Etude de barrages sur modèles réduits en mortiers de plâtre	Colloque International sur les Modèles réduits de Structure, Madrid, 1959
Beaujoint, N., Jacquesson, R.	Emploi conjugé de deux petits modèles plans pour le calcul des micro-déplacements	RILEM Bulletin Nr. 10, 1961
Beeby, A. W.	Short-Term Deformations of Reinforced Concrete Members	Cement and Concrete Association, 1968
Benito, C.	Comprobación experimental de cubiertas laminares, por medio de modelos reducidos	Laboratorio Central de Ensayo de Materiales de Construcción, Publicación No. 97, 1959
Biggs, J. M., Hansen, H. J.	Techniques d'emploi de modèles dans la recherche sur le comportement de structure	RILEM Bulletin Nr. 10, 1961

Author	Title	Publisher/Publication
Bigret, R.	Etude sur maquette des fréquences propres de flexion des poutres	Société Rateau
Borges, F. J.	Statistical Theories of Structural Similitude	RILEM-Bulletin Nr. 7, 1960
Brotchie, J. F., Jacobson, A., Okubo, S.	Effect of Membrane Action on Slab Behaviour	U.S. Naval Civil Engineering Laboratory, California, 1965
Carr, D. P.	Investigation of the Effect of Filler Materials upon the Poisson's Ratio of Polymers	University of Surrey
Charlton, T. M.	Model Analysis of Structures	E. & F. N. Spon Limited, London 1954
Cowan, J. H., Gero, J. S., Ding, G. P., Muncey, R. W.	Models in Architecture	Elsevier Publishing Company, 1968
Desai, C. S., Abel, J. F.	Introduction to the Finite Element Method	Van Nostrand Reinhold, 1972
Eidg. Versuchs- und Forschungsanlage, Emmen	Expo-Signete	Eidg. Flugzeugwerke, Emmen, Bericht 731, 1964
Fumagalli, E.	Matériaux pour modèles réduits et installations de charge	
Goodier, J. N., Thomson, W. T.	Applicability of Similarity Principles to Structural Models	Cornell University, 1944
Gohlke, W.	Mechanisch-elektrische Meßtechnik	Carl Hanser Verlag, München
Gomperts, R., d'Hoop, H., Vichnevetsky, R., Witsenhausen, H.	Anleitung zum Umgang mit P.A.C.E.-Analogrechenanlagen	Europ. Analogrechenzentrum, Brüssel
Grave, H. F.	Elektrische Messung nicht-elektronischer Großen	Akademische Verlagsgesellschaft, Frankfurt a.M., 1965
Hofacker, K.	Elastisch eingespanntes Talsperrengewölbe	Aus IVBH-Schlußbericht, 2. Kongreß, Oktober 1936, Berlin-München
Homberg, H., Marx, W. R.	Schiefe Stäbe und Platten	Werner Verlag GmbH, Düsseldorf, 1958

Author	Title	Publisher/Publication
Hossdorf, H.	Eine programmgesteuerte, voll-automatische Modellmeß- und Datenauswertungsanlage	Schweiz. Bauzeitung, Nr. 39, 1965
Hossdorf, H.	Câble et dispositif de mise en tension pour précontraindre les modèles réduits en mortier	RILEM Bulletin Nr. 10, 1961
Hossdorf, H.	Design of a Polyster Pavilion Reinforced with Glass Fibre for the 64 Swiss Exhibition	IASS-Bulletin Nr. 19
Jones, L. L.	Tests on a One-Sixth Scale Model of a Hyperbolic Paraboloid Umbrella Shell Roof	Cement and Concrete Association, 1961
Laboratoires du bâtiment et des travaux publics	Compte rendu	Centre expérimental de recherches et d'études du bâtiment et des travaux publics
Langhaar, H. L.	Dimensional Analysis and Theory of Models	John Wiley & Sons, Inc., New York, 1962
Lardy, P.	Beiträge zu ausgewählten Problemen des Massivbaues	Verlag Leeman, Zürich, 1961
Le Corbusier	Le modulor, no. 2	Architecture d'aujourd'hui, Boulogne, 1955
Leonhardt, F., Andrä, W.	A Simple Method to Draw Influence Lines for Slabs	"Der Bauingenieur", Nr. 11, 1958
de Leiris, H.	Emploi de moulages plastiques pour le relevé des déformations sous charge	RILEM Bulletin Nr. 10, 1961
Litle, W. A.	Reliability of Shell Buckling Predictions	Massachusetts Institute of Technology, 1964
Litle, W. A.	Structural Models—Some Examples	Massachusetts Institute of Technology
Litle, W. A., Paparoni, M.	Size Effect in Small-Scale Models of Reinforced Concrete Beams	American Concrete Institute, 1966
Little, G., Rowe, R. E.	Load Distribution in Multi-Webbed Bridge Structures from Tests on Plastic Models	Magazine of Concrete Research, Nr. 21, 1955

Author	Title	Publisher/Publication
Mark, R.	Photoelastic Analysis of Thin Plate and Shell Building Structures by the Stress Freezing Method	Princeton University, 1965
Mark, R., Jonash, R. S.	Wind Loading on Gothic Structure	Princeton University, 1970
Miller, C. L.	Man-Machine Communications in Civil Engineering	Massachusetts Institute of Technology, T 63-3, 1963
Müller, R. K.	Ein Beitrag zur Dehnungs-messung an Kunstharzmodellen	Paul Jllg, Stuttgart, 1964
Müller, R. K.	Möglichkeiten und Grenzen der Modellstatik	Aus der Festschrift zum 70. Geburtstag von Prof. Dr. Ing. A. Mehmel
Nervi, L. P.	Bauten und Projekte	Verlag Arthur Niggli, Teufen/AR, 1957
Oberti, G.	Large Scale Model Testing of Structures Outside the Elastic Limit	
Oberti, G.	Rapport général	RILEM Bulletin Nr. 10, 1961
Philips Ltd.	Guide to Strain Gauges	
Preece, B. W., Davies, J. D.	Models for Structural Concrete	CR Books Limited, London, 1964
Porcheron, Y.	Essais de pylones modèles réduits	Colloque International sur Modèles Réduits de Structures, 1959
Raven, F. H.	Automatic Control Engineering	McGraw-Hill Book Company Ltd.
Rocha, M.	Structural Model Techniques Some Recent Developments	Laboratorio Nacional de Engenharia Civil, 1965
Rocha, M.	Dimensionamento Experimental das Estruturas	Laboratorio Nacional de Engenharia Civil, 1952
Rocha, M., Serafim, J. L., Esteves Ferreira, M. J.	The Determination of the Safety Factor of Arch Dams by Means of Models	RILEM Bulletin Nr. 10, 1961
Rowe, R. E.	Shell Research	North-Holland Publishing Company, 1961
Rowe, R. E.	Model Testing—a Realistic Approach for Structures	Cement and Concrete Association
Scott, N. R.	Electronic Computer Technology	McGraw-Hill Book Company Inc., New York, 1970
Serafim, J. L., Cruz Azevedo, M.	Methods in Use at the LNEC for the Stress Analysis in Models of Dams	LNEC Technical Paper No. 201, 1963
Serafim, J. L., Poole da Costa, J.	Methods and Materials for the Study of the Weight Stresses in Dams by Means of Models	RILEM Bulletin Nr. 10, 1961

Author	Title	Publisher/Publication
Somerville, G., Roll, F., Caldwell, J. A. D.	Tests on a One-Twelfth Scale Model of the Mancunian Way	Cement and Concrete Association, 1965
Steckner, S.	Einspannung von Stahlstützen in Stahlbetonkonstruktionen durch Haftung	"Die Bautechnik" Nr. 10, 1969
Stüssi, F.	Der Baumeister Johann Ulrich Grubenmann und seine Zeit	Industrielle Organisation Zürich, 1961
Tsuboi Yoshikatsu, Mamoru Kawaguchi	Design Problems of a Suspension Roof Structure	University of Tokio, 1965
VDI-Berichte	Experimentelle Spannungs-analyse	Verein Deutscher Ingenieure, Düsseldorf, 1966
Waltking, F. W.	Die neue Köln—Mühlheimer Brücke	Herausgegeben von der Stadt Köln, 1951
Whitbread, R. E., Scruton, C., Charlton, T. M.	An Aerodynamic Investigation for the 437 Ft Tower Block, London	National Physical Laboratory, Aerodynamics Division, NPL Aero Report 1032, 1962
White, R. N.	Small Scale Direct Model of Reinforced and Prestressed Concrete Structures	National Science Foundation Grant GP-2622, 1966, Cornell University
Wiss, Janney, Elstner and Associates	Model Tests of Waste Disposal Tanks	1969
Wiss, Janney, Elstner and Associates	Model Studies of TWA Overhaul Hangar for Aero-Shell	1969
Zienkiewicz, O. C., Cheung, Y. K.	The Finite Element Method in Structural and Continuum Mechanics	McGraw-Hill Publishing Company Ltd., New York, 1968
	Die Brücke über den Maracaibo-See in Venezuela	Bauverlag GmbH, Wiesbaden,
	DATALINK—A User's Manual	Massachusetts Institute of Technology, 1965
	Plastics in Building Structures	Conference in London, 14.–16.6.1965, Pergamon Press